KN PRESS - T.R. GRADY

Aphyosemion lugens HAH 88-313
Photo: Richard Pierce

The Encyclopedia of Killifish

Colume I

Basic Care & Breeding

T.R. Grady

Photography
Richard Pierce
Mike Jacobs
Tom Grady
Stefano Valdesalici
Anthony Terceira

KN Press, USA

KN PRESS - T.R. GRADY

Published by KN Press
Nightmist Publishing
27 Webster St,
Malone, New York 12953

The information contained within this book is based on the personal experiences of the author.

ISBN-13: 978-1515106401
ISBN-10: 1515106403
Printed in the United States of America

I would like to extend my personal thanks to those who provided a great deal of assistance in a wide variety of ways.

Dr. Dan Nielsen   Fred Behrmann   Leslie Dick
All the members of Upstate New York Killifish Association
and especially to my editorial staff.
Kim Marshall   Phyllis Grady

Cover Photos:
Rivulus xiphidius - Richard Pierce
Aphtosemion gabunense gabunense - Mike Jacobs
Pterolebias phasianus - Richard Pierce
Title Page Photo: Fp. avichang by Mike Jacobs
Back Page Photo: Aphyosemion lugens JAJ 98-313 - Richard Pierce
Title Page Photo: Fundulopanchax avichang by Mike Jacobs

Foreword

I was about 12 years old when I caught my first pair of fish in the Old Mill Pond about 100 yards from my home. I appropriated my mother's metal potato strainer and a large bucket and headed for one of those sandy spots along the shoreline where I always saw lots of minnows. I figured it would be easy to catch them since I could walk into the water and the fish would swarm around my feet. Not so, but I did end up with my first few fish that I would recognize as Black-line Dace today.

I took my prized catch back to the house and dumped them into an old one-gallon gold fish bowl my grandmother gave me and began to watch them. I added dirt from the back yard and of course it completely clouded the bowl, so the only time I saw the fish was when they came close to the front. I don't know why now, but back then I thought it was cool and I decided I needed a larger viewing platform.

After a little consideration, I found a cardboard box - one of those that hold 4 six-packs of beer - and cut out a small viewing window in the center. I pushed the gold fish bowl tight against the window and stared for hours at that small 4x4 opening. I was hooked on keeping fish.

That Christmas, my parents, recognizing my obsession, bought me my first ten-gallon fish tank, an old Metaframe tank and that just cemented my new addiction. One of the local postal carrriers, Bill Lawrence, found out about my interest and invited me to come to his home and pick out some guppies. Imagine my excitement when the first babies appeared in my tank.

The second tank came soon after and then a trip 50 miles (mind you this is 1965) in an old Rambler station wagon to the nearest place that had tropical fish, a Woolworths. My excitement grew by leaps and bounds upon seeing all those rare tropical fish and we came home with some neon tetras, glow light tetras, a cory catfish and who knows what else.

My hobby grew throughout high school and I ended up with a number of tanks including a 29-gallon high which was inhabited by Elmer the Eel. Yes, my mother named the fish, just like all our other pets that ranged from monkeys, to a goat, to my brothers numerous reptiles and any number of dogs, cats and birds. We were The Menagerie.

How does this relate to killifish?

In the 1960s, the Aquarium Stock Company (AquaStock) put out a catalog. Mind you, I lived 300 miles from New York City. Somehow we were able to get a copy of the catalog and my mother allowed me to order one fish from the catalog. I was drawn to the 'Lyretail' *Aphyosemion australe* and we ordered it. Somehow my mother made arrangements with AquaStock to put the box with the fish onto a Greyhound Bus and shipped to my hometown. At 5:00 PM on a Thursday afternoon, I sat outside the bus stop, actually a small pharmacy/lunch counter and the bus finally came days late (actually only a few minutes). My box was there and I opened it. Where was my "lyretail"? There was a note inside. "Sorry, we had to substitute" and I received a a 'hockeystick' tetra instead.

My disappointment was overwhelming, but it is probably the single most important moment of

- 5 -

KN PRESS - T.R. GRADY

hobby hobby life. From that day forward, I had to have a 'lyretail.' I was truly addicted to killifish and did not even know it yet.

At some point in the next couple of years I received William T. Innes' book, Exotic Tropical Fish and saw a color picture of *Aphyosemion sjoestedti* - a psychodelic fish (remember this is the '60s) and I just had to try to find that one too. (Yes I know it is *Fundulopanchax sjoestedti* today)

In 1972, I joined the U.S. Navy and was stationed in Boston. This was a time of huge growth in my hobby in general and I had about 20 tanks in a small apartment. I still wanted that psychodelic fish. For months I searched every pet store in the surrounding region (probably 20-30 stores) every weekend for that fish and finally someone told me it was a killifish. A what? I was pointed to a membership advertisement in TFH magazine and I sent in my $10 to join the American Killifish Association. I looked in the Fish & Eggs Listings, was excited to find *A. sjoestedti* listed and ordered my psychodelic fish. Once again disappointment struck. I didn't want a Blue Gularis, I wanted that misidentified psychodelic fish, which today we know is actually *Callopanchax occidentalis*.

Yep, the obsession grew.

Finally in 1975, the A.K.A. was holding its annual convention on Long Island. I gathered up the wife, the baby and off we went. I was awestruck. Some incredible people, Chris Baker-Carr and Charlie Nunziata made me feel very welcome and I have not turned back from this hobby of killifish since. One fish, *Nothobranchius kirki*, an incredible almost completely red male, was the center of attention. It may have simply been a sport, but to this day I have never seen another like it. That was the fish that brought my focus to Nothos, still my favorite group of killifish. Stuart Grant of Malawi was the guest speaker and we maintained contact until his passing several years ago. I sat down with some old guy from Chicago and we spoke for hours about killies. That old guy, whom I did not know, is a true legend of this hobby - George Maier. I had no idea whatsoever. However, I met some life long friends at that convention, people I still look forward to seeing. I spend quite a bit of time traveling to conventions together and sharing vacations even now 40 years later, Dr. Dan Nielsen. We are both members of Upstate New York Killifish Association.

I think back to just how long I have kept tropical fish, over 50 years now, and the last 40 were in killies in one way or another. I have been privileged to hold office on the board of the AKA and been actively involved as a committee chairman for many years. I worked on the AKA Website and was the complier of the monthly business news letter for many years. I have been fortunate to be recognized as Killifish Hobbyist of the Year and as a Distinguished Member of the AKA. For many years I was chairman of the AKA Convention Assistance Committee and I also chaired or co-chaired five national conventions. I have been actively involved with Northeast Weekend, a regional mini-convention, since its inception and traveled to other regional shows. Everywhere I go, I find new friends. In many ways, this hobby opened up the world for me. I have collected killies in the Caribbean and many areas of the United States and still hope to someday go to Africa or South America.

This hobby has offered me friendships around the world, the opportunity to collect fish and a chance travel to places I would never have seen otherwise. In a large part, that is why I chose to write **The Encyclopedia of Killifish** - to offer my thanks to all those who have contributed to my love of killies and the try to give something to the next generation of aquarists. This series of books is designed with the beginner and intermediate hobbyist in mind with some tips, anecdotes and plain out warnings that hopefully will be of value to all levels of killie fanciers. Enjoy!

Tom Grady

Table of Contents

KN PRESS - T.R. GRADY

A. bitaeniatum Ijebu Ode   Photo: Mike Jacobs

Just What Are Killifish?

Where are Killifish Found

With the exception of Antarctica and Australia (where killies are ecologically replaced by Rainbow fishes), killifish thrive on every continent around the world. They can be found in small streams and large rivers, in puddles of residual rain and the largest lakes and even in some of the most inhospitable of conditions. Pupfish, *Cyprinodon* species, are found in the extreme heat of America's Death Valley and on a narrow shelf of a deep cavern (Devil's Hole in Nevada) where perhaps less than one hundred of these nearly extinct and heavily protected fish exist.

It is possible for a fish keeper to go to a backyard stream nearly anywhere in the United States, local ditches from Spain to Greece or a nearly dry pond in Africa and catch one of the many astoundingly beautiful small species that inhabit those bio-types. Killies have been used as bait fish as well as, and to the horror of hobbyists, are considered a food delicacy in some parts of the world. When Professor Wilhelm Peters first described *Nothobranchius*, he noted the local peoples of Mozambique consider them as delicacies and claimed "When baked they provide an excellent dish". (R.A. Jubb - Nothobranchius, pg.8)

In some cases killifish have been introduced, with limited success, as a means for mosquito control and one species has actually traveled into outer space.

These amazing animals have evolved survival mechanisms that allow the species to survive extremes of environment ranging from the complete evaporation of the waters they inhabit to the ever changing salinity of coastal tidal zones where water can change from fresh to salt in a matter of minutes. Within the same ecosystems, killies can be both the hunter and the prey, the carnivorous predator and its smaller victim. Certain species have developed unique behaviors such as the scale eating *Cyprinodon desquamator* on the Caribbean island of San Salvador and the internal insemination of *Campellolebias brucei*. In East Africa, members of the *Nothobranchius* genus lay their

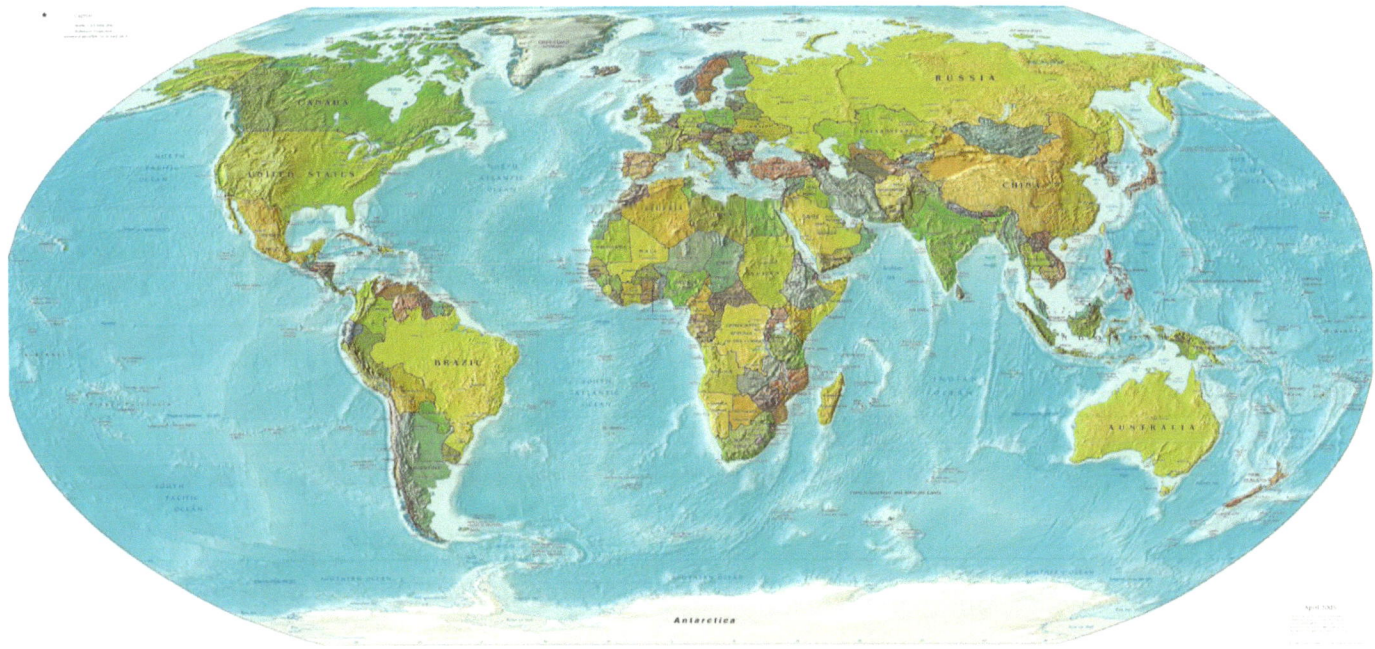

eggs in the bottom substrate and then die off when the water evaporates and the land becomes arid. Months, in some cases even years later, the rains return to fill up the pools and ponds and the eggs hatch into a new generation. In South America species of the *Rivulus* genus leap from the waters and cling to the vegetation above to escape their enemies and in some cases, certain species actually lay their eggs at the water line. In some parts of Death Valley in North America, the water temperature can reach 110F (43C) and in Argentina the surface of ponds can freeze, yet the fish survive, even thrive.

A large part of the draw for killies as aquarium fish is the challenge of providing an environment that allows for the successful reproduction. Hobbyists may originally be drawn to the brilliant colors, the unique behaviors and the active personalities of killifish, but as they begin to experience success in breeding some of the easier species, the draw of attempting more difficult members can lead to an almost obsessive desire to accomplish what few have achieved. Killies can provide a true adventure for every level of fish-keeper.

In addition to maintenance in the home aquarium, killies often leads a hobbyist in another direction, that of collecting. The vast majority of species in today's home aquariums have been initially caught by hobbyists, not professional or scientific collectors. Aquarists from around the world travel to unexplored regions in hopes of uncovering new species and it happens on a regular basis. Descriptions of previously undiscovered killies are announced nearly monthly and come from nearly everywhere on the globe. Perhaps the largest number over the past few years have come out of South America, but Africa is not far behind as formerly dangerous areas become safer for exploration.

Admittedly, hobbyists may be the primary group to seek out new fish, but it is still taxon-

omists who give them status. It is interesting to note, despite criticism, even the inexperienced hobbyist can also become the taxonomist. Many species of killifish have been named and described by hobbyists. In fact there is a competing subculture of scientists and hobbyists who constantly challenge each other's work. In a sense, no matter who is right, the popular fish hobbyists have come to know very well as one species can suddenly be named something completely different when another taxonomist decides a change is necessary for some abstract (valid or not) reason.

For hobbyists this can be a real problem with identification and often the same species is known by two or three different scientific names and it takes some real effort to remain current. Yet 'common names' are not popular with killifish enthusiasts. In part it is too easy for a common name in one place to be completely different elsewhere.

An interesting side-note on the naming of killifish is the strong desire of most killie hobbyists to actually want to know where the fish come from even to the population level - exactly what pond or pool of water near what village, GPS location or collection code designation where the fish were originally collected.

Fundulopanchax sjoestedti **Dwarf Red** *Gularis*
Photo: Mike Jacobs

Scientifically this is important information that sometimes leads to a taxonomic separation of similar species.

The Scientific Side

Killifish are the egg-laying toothcarps of the Order *Cyprinodontiformes*, however, livebearers (*Poeciliidae*) are also members. Further evaluation reveals a large group of *Poeciliidae* (species identified as Lampeyes for example) are actually oviparous (egg-laying) species. After that it becomes more complicated. There are a number of written opinions on the relationship of killifish and livebearers. Most agree in one way or another. In reality it is the terms for the different levels that seem to have the most disagreement. Two suborders are recognized by most authors, *Aplocheiloidei* (Bleeker, 1859) and *Cyprodontoidei* (Wagner 1828). These suborders are further split into a number of families. The divisions primarily represent killies of the eastern and western hemispheres and then further denote other regions (North America vs South America). They are further separated into a variety of tribes, subtribes, genus, subgenus, and subfamilies depending on which line of thought followed. For example, in 2000 Dr. Jean Huber created Tribes and Subtribes beneath the Family level before moving to Genus status. In essence this was an extension of an earlier effort in *Nothobranchius* by Garman in 1895. While Dr. Huber's work is noted, it is not necessary to the hobby level discussion.

For the taxonomist, these distinctions are important, but for the hobbyist, it becomes a little over the top. For the most part hobbyists are interested in the 'correct' name of a species they are keeping, not the systematic placement of the fish in various theories. Despite the various genus and/or subgenus names, without a doubt the killie-keeper has a tendency to use terms defined from common usage instead of the taxonomic. A good example of this would be the common usage of *Fundulopanchax* instead of the subgroup name *Gularopanchax*. Even within the hobbyist culture there is often disagreement on the 'proper' name of many killies. Most people prefer the use of genus names over subgenus. A large part of the reason is when the fish were originally obtained, they came with the genus identifier. At one time all *Rivulus* were simply *Rivulus,* but today this genus is broken down into ten subgenuses.

Because taxonomists break the species down into sub-genus status, this creates a con-

***Aphyosemion splendopleure* Njombe**
Photo: Richard Pierce

***Aphyosemion australe* BSWG 97-24 Port Gentil**
Photo: Mike Jacobs

stant state of flux. It sometimes takes years for the changes to become settled, but sooner or later even the aquarist comes to know the fish at the subgenus level. This issue is not restricted to killifish, but runs rampant throughout all of the major groups of fish from cichlids to tetras.

In order to make this work viable for all levels of interest, systematic chart is is included in **Appendix A** to depict current thought on the taxonomic status of killifish. This list is not meant to challenge the taxonomy of killies, but instead represents a working concept of how hobbyists can view the fish they keep in their

---

aquariums.

As with everything within the killifish community it is likely to become outdated as soon as it is printed. Future volumes devoted to specific groups of killifish will take a more involved look at the theoretic subdivisions.

Why exactly are Latin names used?

The answer to this is simple. Common names can often mislead hobbyists as to what species or potentially even what genus a fish belongs. Too often names like Lyretail or Panchax are used for a variety of species and miscommunication can lead to frustration. By using the scientific names of killies (and for that matter any group of tropical fish) everyone should be on the same page.

Realistically, learning the names of killifish is not at all difficult. Latin names are phonetically spoken and with just a little exposure become part of the international language of the hobby. In essence, everyone from Japan to Britain use the same name to discuss the same species.

The system created by Carl Lannaeus is used for all organisms, plant or animal or otherwise. Technically it is a binomial system, meaning two names. The first name identifies the genus, and the second the species. While it may initially be a little disconcerting, it does not take long to become comfortable with these names. No one needs to learn the dead language of Latin unless they have a real desire to

---

How it works:

Genus Name	Species/Subspecies		Population/location
Fundulopanchax	gardneri	nigerianus	Makurdi (A village in Nigeria)
Nothobranchius	rubripinnis		Mbezi River TZ 83-5 (Tanzania 1983 5th collection stop)

---

understand exacly what the names denote. Instead to simply assimilate the names as part of whatever language makes it much easier. This way the terms develop common usage across all languages, be it English or Chinese, French or Korean.

Because there are over 1,100 species of killies and a large number of populations, common usage often includes the original collection location of the fish in a couple of

different ways. Often there is an identification of a specific town, village, river or place where the fish originates. Additionally, collectors use codes to identify locations that do not necessarily have names, but are recorded as geographical coordinates or as some sort of personal record.

In common use, most hobbyists normally identify the first as *Fp. gardneri* Makurdi when discussing the fish and the second would most likely be called *N. rubripinnis* 83-5 or *N. rubripinnis* Mbezi in conversation. Because the identifiers are specific, most hobbyists come to know exactly what fish those are.

Another very important point - at the population level - identification keeps similar species from being hybridized. In some cases they may just become an aquarium strain, but in others two different species may produce a viable cross and may over time be unable to survive.

Hybridization across populations is a somewhat controversial issue within the hobby. Purists insist that no populations should be crossed, in part to maintain the integrity of the collection as well as personal desires to breed specific species. Others feel that crossing populations actually strengthens the species. Both sides present good arguments and valid points cane be made for either stance.

It's interesting to note how many species have developed usage names, simply by using the species name such as '**Guentheri**'. Every experienced killifish hobbyist knows this is *Nothobranchius guentheri* despite 'guentheri' being used widely across all genera of organisms. There are many examples of common terms that have developed over time. The fact is, the longer the species has been in the hobby, the more likely it is that fish will be commonly known by its last name.

Are Killifish Good Community Fish?

The answer to this question is an unqualified ... maybe. Like most larger groups of aquarium fish there are some species which are perfectly fine as a community member and others that will enjoy the various flavors of their tank mates. In nature, some killies are predators, others no more vicious than a Neon Tetra. The behavior of killifish can also a qualifying issue. Some species are suited to a mixed environment, but others prefer a more solitary one.

Many members of the *Aphyosemion* groups are gentle and add intense coloration to any aquarium environment. *Aphyosemion australe* - commonly known as the Lyretail - in its orange form is an incredibly beautiful small fish that will get along fine in a community of tetras, corydoras and other popular fish. But it's larger cousin the Blue Gularis (*Fundulopanchax sjoestedti*) will relish any smaller fish it can fit into its mouth.

On the positive side. Lampeyes provide a striking shoal of fish that actively swim near the surface of the tank. The reflective blue eyes and shining sides of many members of this group of fish add to the community as they constantly move around the tank. But on the other side, members of the *Diapteron* grouping are small fish which move with an intent, slowly working from one spot to another in the tank as they search for the tiniest morsel of food or for a member of the opposite gender with which to reproduce. Small fish of this type, no matter how beautiful, do not do well in an community environment. Faster, more active fish will beat them to the food, while even slightly more aggressive species will find them easy victims to bully or worse the perfect meal.

Other beautiful killies such as *Fundulopanchax gardneri* make for decent and attractive community fish. There are a number of populations of '**Gardner's Killifish**' available, each individually distinctive, with blue sides covered in red spots and variations of yellow, orange and white bands in the fins. Their coloration makes them a beautiful addition. Generally this group of fish is not aggressive, but sometimes individuals can become bullies. If they are in a tank of active, fast moving fish, there is little likelihood of problems.

In a slightly different concept, the creation of a killifish community tank can be rewarding. A 30-gallon tank filled with one species or another can be fascinating. An environment of this type can become self-sustaining. Under the right conditions, including enough 'hiding' places such as plants, rocks and driftwood, the killies will reproduce and maintain their own community. A setup of this nature can sustain itself for an unknown number of generations

mix similar or related species in case of hybridization. But also most females tend to be similar and the hobbyist may not be able to determine which female goes with what male. If an aquarist intends to breed a species, mak-

Aphyosemion australe orange

Photo: Richard Pierce

with just some basic care (water changes and feeding).

The hobbyist will find all ages of fish contained in this system. It is interesting to see some of the dynamics of this type of community. As pairs are removed for sale or die off, new fry will appear to replace them. The author has maintained a 30-gallon tank of *Rivulus frommi* PAN 09-23 for many years after initiating it with a single pair. The population stabilizes at about 20 fish with a variety of ages and sizes.

There are any number of species of killifish which will live comfortably with each other in a larger system. The primary concern is not to

ing certain the pairs are matched becomes a concern.

On the other hand, because of the behavior of killifish, the stop and go movement, there are limitations to what the other residents of a community tank should be.

First, will the other fish beat the killies to the food? This is a problem for many killifish. Rapidly moving adn agressive tropical fish will eat most food up before the killies have a chance to aobtain enough nourishment. Killies have a tendency to seek out their food and not attack it. Many if not most killifish are ambush predators, hiding in the plants and waiting for an

Introduction To Care

opportunistice attack. Fast moving fish simply get to the food first. So killies need to be with other species that will not dominate the tank.

In addition, the community should not contain any aggressive species. Again, killifish tend to be slow moving and can easily become victims for territorial species. While a few species of dwarf cichlids get along, when a pair breeds, it protects its eggs and will charge anything that comes near. Killifish are susceptable to attacks.

The diversity of killies, the varied habitats and the different techniques to successfully propagate different groups and even individual species offers a challenge for the hobbyist. With over 1100 species now described, killie keepers have the opportunity to experiment with new or unique techniques to breed and maintain their favorite fish.

Killies can be found in nearly every niche in nature. **Lampeyes** are mid-to-top level schooling fish while certain pupfish are algae grazing species tending to live in hot springs. Certain members of both African and South American annuals are serious ambush predators while their prey are small, rapidly reproducing annuals in the same pools, ponds and puddles of water. Those same annual species live environmentally shortened lives, sometimes as short as six to eight months in nature. In West Africa many species of the *Aphyosemion* group live in the dense vegetation along the edges of streams and brooks and this is duplicated by *Rivulus* in South and Central America.

Water conditions vary considerably also. Some North American species live in the hard brackish and even salt water along the coasts while others are found in soft water swamps. At times, particularly for the rain forest species, water conditions can change in an instant when heavy rains fill the streams to overflow-

ing. The temperature of the water can go from warm to cold in seconds, whatever nutrients (or man-made chemicals for that matter) in the soils will be washed into the creeks. Killies need to adapt on the fly. In some areas of the world, as the water evaporates, the decaying of plants cause conditions to change vastly over short periods. What was once a lush body of water filled with plants and reeds becomes parched landscape. Yet through the mechanism of diapause **(see page 68)** the species survive.

In Argentina and Uruguay annualism has taken on a slightly different aspect. Many of these killies prefer cooler water conditions, become far more active in the temperature range of 55F-65F (12C-18C) and become listless at the room temperatures in a normally maintained fish room. The ponds may still dry out, but in other cases the waters recede, but remain in much smaller areas. Breeding takes place along the original shoreline which does become dry. The breeder has to find ways to recreate conditions in the home aquaria to allow for these fish to successfully reproduce. Admittedly, some species simply need water, but others require far more effort and as natural duplication as possible to provide an opportunity for success. Hobbyists would be wise to study the natural conditions a particular species lives in in nature. While this is not an absolute, there is little doubt the greatest potential for successful reproduction is replication of the wild conditions.

What this all leads to is an incredible diversity of species and a wide variety of techniques to afford the hobbyists an opportunity to face the challenges of killifish. Each group of fish offer their own requirements and/or options for reproduction, yet within a group, individual species can require something just different enough for the breeder to seek a distinctive 'key' for success. The choice may be water pH or hardness, or it may be something as simple as the color of the spawning mops. Some species may be seasonal breeders while others only seek a partner anytime or anyplace.

While there is a great deal of information about most species of killifish, there are still real challenges out there - species not successful maintained over time or even bred as of yet.

The Killifish Home Environment

The first thing we need to replicate is an environment in which to successfully maintain killies is the living quarters. Obviously, hobby-

> **Personal Experience:** The author has maintained a 30-gallon community tank of Rivulus frommi PAN 09-23 since 2009. The system maintains a balanced group of approximately 20 fish all the time. It is interesting to note the species reproduces in a seasonal manner, producing fry in the Spring every year, but at no other noticeable time.

standard glass tanks are by no means the only option. Because killifish are incredibly adaptable, the vast majority of species will survive in extreme conditions, even in the home aquarium. One breeder was known for maintaining his entire breeding population in 1-gallon pickle jars, while another used semi-opaque sweater boxes. There are as many options as there are hobbyists, but for the purposes of this book, the focus will be on the more traditional method of fish tanks and some options on how to maintain populations of killies.

There are several opinions on what size tanks should be used. The ten-gallon tank is very popular and its uses are multiple, ranging from breeding tanks to raising and holding tanks. Five-gallon tanks are commonly used for breeding pairs and in some cases 2-1/2 gallon tanks are used. Some breeders use a different technique with larger tanks (20-30 gallons)and try to create an environment where the fish reproduce as a community and those systems produce various stages of growth from new born fry to mature fish.

Perhaps the main question each person has to answer is what they want to accomplish within the killie hobby. Does the individual want to just maintain some pretty fish or do they want to face the challenge of breeding them? It is not uncommon for someone who

> **Note:** One unmarried friend of the author had tanks in every room of his multi-bedroom apartment including several in his own bedroom, on the kitchen counter and even in the bathroom.

successfully breeds one species to want to move on to another. It can become an obsession for some and their hobby expands exponentially from that single tank to several. It grows and then more tanks take over the living room and finally the largest available space such as a basement or garage becomes filled with tanks, and work spaces. The need to add tanks becomes overwhelming and the hobbyist ends up with literally hundreds of tanks, pickle jars and sweater boxes housing thousands of fish.

In the end, the size of fish tanks is up to each individual and what they want to accomplish. Even how the tanks are displayed becomes one of choice, space over numbers, or viewing the fish over providing more room to maintain additional species. Many serious breeders display the tanks sideways instead of front-on simply to fit more tanks into a smaller space.

What this mix of information really means is that prospective killie-keepers have a huge number of options on how they house their fish, what choices fit their lifestyles and even attempt to keep spouses from throwing them out of house and home.

A Basic Setup

Without a doubt the ten-gallon (37.5L) tank is the most commonly used housing in the tropical fish hobby and it is popular with killie hobbyists as well. A filtration system and aeration are really all that are necessary, although many species do well in a completely bare tank. The simplest system is a box filter fed by a small air pump. Over the years numerous types of filtration media have been used, but the most basic is aquarium gravel added to a box filter and then covered with polyester filter floss. (A discussion on a variety of filtration options begins on page 25).

Now that the basic tank is set up, it is time to consider whether or not to use aquarium gravel, add live plants or use plastic, and make choices as to what is added to the tank and why.

Stands & Shelving

The number of options available for displaying (or not displaying) killifish are as varied as the hobbyists keeping them and so are the possible negatives. Should the tanks be displayed front-side out or sideways? How many tanks does it take to fill a space and what are the base sizes of tanks in order to plan a layout? How much wall space is available, how much open floor space? More importantly, what is the best design to maximize the usage of the fish space?

At that point, the aquarist needs to know how much weight the stands can bear and whether or not the hobbyist intends to expand in the future?

There are any number of shelving units that can be purchased commercially. Most are not designed specifically for the weight of aquariums, so care must be taken in selecting them. However, several do make for solid stands. Most are not going to match the dimensions

In this case, the use of pine 2x4s are braced every 4 feet and can hold a great deal of weight. In the center of the photo, there is a 2x2 beam placed beneath the 2x4 to hold the highest section level. This articular rack holds 24 ten-gallon tanks with a filled weight of over 1,900 lbs

of the fish tanks either, but a few have enough leeway to be of value. Just be absolutely certain the stand can handle the weight of however many tanks are placed on the shelves.

Of course, pet stores sell very expensive individual stands that might hold two tanks. In the name of savings, these are only recommended for display tanks in the public areas of the household.

Home-Made Stands

The use of homemade stands is common in the serious aquarium hobby. These can vary from 2x4 framed stands to planks placed across and stacked on several levels of cinder blocks. Of all the designs fish keepers use, one of the simplest, yet most useful for full length frontal displays is a shelving unit constructed with 2 x 8 (for 5-gallon tanks) and/or 2 x 10 (for 10-gallon tanks) heavy planks.

The completed unit is attached to the wall with 'L' brackets to maintain stability. The length of the planks should be measured to provide a small amount of space between each tank although the tanks can be flush if that is all space will allow. The depth of the boards should match the width of the tanks (8-inches for 5s, 10-inches for 10s, etc.).

The end boards - the weight bearing component - should be the same as the widest tank. Drywall screws may be used to put this together. Alternatively 2x4s can be used on the ends. The separation of the shelves should leave at least four to six inches above the tanks for ease of access. If the end boards are slightly narrower than the shelves, or offset by an inch, the design can incorporate a space to allow air tubes and extension cords space to pass behind and above the tanks

For smaller 2 1/2-gallon tanks all that is needed is a one-inch thick plank of 4 or 8 inches width depending on how the hobbyist wants to view the tank. The length is whatever is needed. These can easily be placed on metal

'L' brackets attached to the studs in the wall. Enough space between each level is needed for access. Make certain the 'L' brackets are screwed into the beams that hold the wallboard or into 2x4 permanent braces.

Compact Units:

For the breeder who wants to fit as many tanks possible into the smallest square footage because of space limitations or just the desire

The use of 2x8 or 2x10 planks allows for less wasted space, particlarly for smaller tanks, although the size will support several 5-gallon tanks with extra support.

to have hundreds of tanks, perhaps the best option is a 2x4 Compact Unit. These racks are designed to fit the maximum number of tanks into the smallest possible space. A four foot long unit that is 16-inches wide can handle up to 24 five-gallon tanks sideways on four shelves. There are various alternatives in design for fish keepers who want to have a mix of tank sizes, but all are constructed in the same way taking into account tank sizes.

The entire frame and all shelves are 2x4 lengths of pine. For 5-gallon tanks each shelf is planned to be four feet long, then the cross beam must be cut to 10 inches. This is put together with 3.5-inch dry wall screws in a rectangle. Once four shelves are created, they may be attached to two end stand frames that reach from the floor to roughly the ceiling. The bottom shelf can be set on the floor and each succeeding one a planned distance above. Be sure to use a level and mark each on the side panels. The separation of the shelves needs to be enough for there to be 4-6 inches separation between the tops of the tanks to the next level for ease of access. Five-gallon tanks are 10-inches high, so the shelf should be leveled 16 inches above the lower row. Keep in mind the 2x4s themselves take up space. Technically this makes each shelf 20 inches from top to bottom. Four shelves then take up 76-inches of height or are about six feet tall. There is a little workable room if necessary with a variance of eight inches if the space between shelves is narrowed to four inches.

However, if a combination of tank sizes is used and the plans modified for fit, then the top level of the stand will be within easy reach. In addition, the hybrid units described below systems also works quite nicely and allows for different size tanks to be contained in the same space. The bottom two shelves could be for 10-gallon tanks and the upper two for 5- and/or 2-1/2 gallon tanks. It is all a matter of what the aquarist needs.

All that really matters is the needs of the hobbyist and how much money one has on hand. Whether you have one or two tanks or a are even a better use of space.

Depending on the space available, these units can be connected together and several aligned along wall areas. In addition, this base design is no different for free-standing units, but those should be attached to ceiling cross-beams for stability if available (such as in a cellar).

Realistically, this style of stand can be designed to hold any size of fish tank. The only consideration is the size and whether additional support is needed to keep the 2x4s from warping.

Several of these racks can be designed for 20-long, 20-high and 30-gallon tanks that hold up to three levels and can be mixed and matched to create unique combinations. They are attached to each other to create stability.

Hybrid Units:

A combination of the first two shelving room of hundreds, it is wise to map out a long-term plan to be ready to take the next step. There are many ways to build tank racks that are different than what is suggested above. Some aquarists use cinder blocks and stack planks across them in a very quick and easy way to create shelves, others create elaborate systems that include the air and filtration systems.

Water Quality: The One Thing That Needs Constant Attention

Water, water, everywhere water, but is it good for your fish?

Probably the single biggest killer of tropical fish, be it killies or anything else, is water quality. There are so many factors ranging from ammonia to nitrates to pollution that affect the health of the animals. The way a hobbyist deals with water must be viewed as the number one priority in maintaining the environment where the fish are kept.

What exactly is meant by water quality?

Water in nature is constantly renewed. Rain falls, fills streams and creeks which flow into larger flows and finally ending up in rivers that connect to a lake or the oceans. In many cases ground water rises in springs that then flow to a final destination. The constant change of water in those habitats essentially means the conditions are 'clean' most of the time. It is very difficult to reproduce the constant change of conditions in nature, but it is important to come as close as possible within these miniature home environments.

What happens in the home aquaria?

New aquariums need to 'cycle'. This actually means is the filtration system needs to catch up with the waste products produced by the fish and plants present in the tanks. This is accomplished via 'good' bacteria that become established in the filter media and gravel (if present). In a new system, this takes 30-45 days if done from scratch. The time can be shortened by using established filters (filters used in another aquarium moved into the new tank). There are also additive bacteria that can be found in some pet shops, but these may not work as well as an established filter. (* See the section on filtration.)

This is known as the Nitrogen Cycle.

There are other methods to control nitrates in aquariums besides water changes. For freshwater fish tanks, live aquarium plants will use up some of the nitrates. In saltwater fish tanks,

---

The Nitrogen Cycle

- **Stage 1**

  Tropical fish waste and uneaten food introduces Ammonia into the aquarium. The food and waste break down into either ionized ammonium (NH4) or un-ionized ammonia (NH3). Ammonium is not harmful for fish, but ammonia is. The pH level of the water determines which product is the end result. Ph under 7.0 will produce ammonium while pH over 7.0 will convert to ammonia.

- **Stage 2**

  Nitrosomonas, a bacteria will develop in the filter media and substrate (if there is gravel or sand in the aquarium). Nitrosomonas oxidize the ammonia. This essentially eliminates the chemical from the system. Nitrites are formed during this process. With the introduction of Nitrites into the system we face a different toxin. Nitrites are just as poisonous to tropical fish as ammonia. Using a test kit, about two weeks into the cycle, the hobbyist will see a rise in Nitrite levels.

- **Stage 3**

  At some point around four weeks, bacteria called nitrobacter will develop. This bacteria will convert the nitrites into nitrates. While not as harmful to fish as ammonia or nitrites, nitrate is still harmful in large does. This is where water changes come into play. The best way to eliminate nitrates is to perform partial water changes. Once your tank is established it is wise to continue to monitor the tank water for high nitrate levels and perform water changes when necessary.

---

live rock and deep sand beds can have an-aerobic areas where denitrifying bacteria can breakdown nitrates into harmless nitrogen gas that escapes through the water surface of the aquarium.

Water Changes

As much as most hobbyists hate the work, the only real way we can create as clean an environment as possible is to do regular water changes. The aquarist has to understand that these small enclosures where the fish live are as much a toilet as a home. Fish are fed a variety of foods ranging from live to frozen to dried flakes. If the food is not eaten, it lies on the bottom and decomposes. When fish eat food, they produce wastes - urine and feces. All of this remains in the container unless we remove it, period. Over time the chemistry will reach disastrous proportions and fish will die. Before they reach that point, breeding behavior tails off, the fish settle into certain locations in the aquarium and often just look bad, scales may stand out and the body becomes bloated, and finally death by poisoning, in its most simple description, occurs.

There are options for the aquarist when it comes to water changes, but reality needs to come into play. A constant water change system can be installed, a series of PVC tubing connected to each tank with a drainage system as well as refill reservoir and lines to replace the water. The effort and costs may be fine for some people with the plumbing skills to create such a system and it certainly is of benefit. (See *http://aka.org/flow-through-system/* on how to build a water change system.)

Most hobbyists simply drain the water from the tanks by hand and then refill them. Nearly every killie-keeper has a method and tricks developed over time. Years ago, the only way to change water was by hand, simply draining the water from a tank into a bucket and then dumping it down the drain. Water was then added bucket by bucket back into the tank. Today there are a number of variations on that age old technique.

Hose System (Python, Aqueon): The hose system removes a great deal of the heavy lifting from the effort. The Python, and variations of it, use pressure from the faucet to draw water from a container and alternatively replace the water into the tank. This can be done in a direct way by simply inserting the hose into the fish tank and draining the water, or a more indirect option by placing a large container (bucket, barrel or plastic garbage can) on the floor to hold water suctioned directly from the tank with a smaller hose by the hobbyist.

The Python hose system removes the water from the large receptacle while the hobbyist uses a small handheld hose to drain each tank into the larger holding container. This certainly can speed up the process.

To refill the tanks, the same Python can be used simply by reversing the flow on the faucet valve. One important thing to remember, it is wise to add a dechlorinator to each tank being refilled if you use tap water and be certain to make sure the water temperature is reasonably close to the tank water.

An alternative to this system to refill tanks is to use a water pump from a holding tank of pretreated water.

One hobbyist created his own 'octopus' system which uses several hoses attached to a central container at the same time to drain the tanks. In essence this is a more elaborate version of the hose system.

Acidity & Alkalinity

The acidity or alkalinity, known as pH (numeric scale) can have a significant affect on tropical fish. Since water is constantly exchanged through osmosis in the cells of the animals, a shift in pH can affect the fish's blood acidity. Fish require energy to regulate internal pH and a rapid shift can cause serious stress and even lead to death. While the shift in pH occurs in the wild and fish adjust in response naturally, the changes are not usually rapid.

Technically the pH stands for **pondus hydrogeni** (the Latin for Potential of Hydrogen). The higher the concentration of hydrogen ions in the water column the lower the acidity of the water becomes. The scale is based on a neutrality of 7.0 and extends in both directions from 0.0 on the acid side to 14.0 alkalinity. Each step (number) in the scale represents a 10-fold exponential increase or decrease of the acidity/alkalinity.

Both acid and base extremes are deadly for fish, although there are a few species which can tolerate levels close to the limits. The preferred range, depending on a specific species, tends to run from 6.0pH to 8.0pH for freshwater fish.

0-4 dGH	(inder 70 ppm)	very soft
4-8 dGH	(70-140ppm)	soft
8-12 dGH	(140-210 ppm)	med hard
12-18 dGH	(210-320 ppm)	fairly hard
18-30 dGH	(320-530 ppm)	hard
^30 dGH	(over 530 ppm)	very hard

Killifish exist in nearly every possible inhabitable pH range in the wild, extending from high alkalinity waters of the oceans to very low acidity of tropical rainforests. Most do adapt well to local waters within the normal ranges, but some species do far better in conditions closer to their natural bio-type. There are always exceptions to every rule, but some killies will not reproduce in anything less than near perfect conditions, be it water acidity, hardness or temperature. Sometimes it is a good idea to learn the conditions of the original habitat from which the fish comes from to be successful.

Today acidity can be easily tested with a wide variety of kits available in hobby stores and the quality of these kits vary, but most will give a reasonable approximation of the actual water condition. Once the pH is determined and it needs to be modified to meet certain requirements of the fish species, there are a variety of chemical products - called buffers - on the commercial market that can be used to raise or lower the acidity.

Alternatively, and probably a better choice, would be to use natural substances to alter the water chemistry. Peat moss, placed in a filter, will lower the pH and calcite chips (or crushed sea shells) will raise the alkalinity over time. The major problem with Peat moss is it will decay in water and release organic compounds called tannins into the water column. The positive side is that tannins may have some health benefits for the aquarium by precipitating certain chemicals from the water. On the negative side, the release of these organic chemicals can color the water to a tea-like appearance. This can be seen in the photo below where the tank has peat moss on the bottom

Be aware that once a hobbyist begins to alter the water acidity, it can lead to an unstable system. While killifish tolerate a wide range of pH, just surviving is not the same as thriving. Generally an aquarist has a reason for altering the water chemistry, so recognition of the potential for rapid spikes and drops of the pH is necessary. Since this can stress the inhabitants, it would be best to avoid large scale changes to the environment.

The other thing that can lead to pH changes is uneaten food decaying on the bottom or in the corners. Bacteria form to eat this left over food and break down the matter, a byproduct of which is the addition of carbon dioxide to the system. CO_2 will lower the pH. The simple answer to this is to remove any organic food, decaying plants or waste products and do water changes.

Water Hardness

Water hardness refers to the amount of dissolved solids suspended in the water column. Hardness plays an important part in the health of fish. Growth along with then passage of nutrients and the elimination of wastes through cell membranes can directly affect the fish. When spawning, it can affect the transfer of sperm and egg fertility. Many of the fishes internal organs, particularly the kidneys, can be impacted.

Two measurements are commonly used in the aquarium hobby,. TDS (Total Dissolved Solids) is a measure of all inorganic and organic substances while GDS (General Hardness) is focused on the minerals calcium and magnesium. Depending on the test kit, different measurements are recognized. The following chart shows the range of hardness for hobby purposes.

In the home aquarium, water hardness is important based on individual species. Some killies prefer harder water, while other do best in softer water. Very often a care examination of where the fish come from in nature will reveal the best options. Most of the rainforest species come from places where the water is very soft, often no more than 4dGH, while many of the Pupfish come from brackish and salt waters with a hardness of 12dGH or much higher.

It would be wise to know what the hardness of the local water system and monitor it every time new water is added to aquariums. There can be radical changes based on the season in many parts of the world. Water companies often add chemicals to change the hardness of drinking water and this can be different from town to town.

It should be noted that rain actually has no hardness in its purest form, but sometimes air pollution can affect it in various ways, so testing it and creating mixtures with tap water based on current needs is very important.

Temperature

Every aquarist is warned about water temperature by local fish store workers when they first set up an aquarium. After all aquarium fish are 'tropical' fish. The hobbyist is often told to float the bags of fish on top of the surface to allow the water temperature to adjust before releasing the fish into the tank. There are far better ways to help the fish adjust to the tank conditions than concern with small temperature changes.

The vast majority of 'tropical' fish actually come from a wide variety of tropical and temperate conditions in nature. As far as temperature, this is pretty much one of the fallacies of the commercial hobby. Reasons to slowly adjust new fish to your water include the previously discussed pH and water hardness.

Consider this: In the wild, fish can be subjected to large temperature changes, particularly in small brooks, streams and ponds when the rain falls. The swelling of rivers with heavy rainforest downpours modifies the temperature rapidly. Add to the fact, larger ponds and lakes contain a thermocline, a temperature barrier of several degrees between water layers. Fish constantly pass through the thermocline without difficulty. While fish are cold blooded, they can regulate their temperature to adjust to the changes so long as the change is not massive.

What the aquarist does not want to do is make a huge change in the temperature conditions if possible. Certainly there can be shock if a change of 50º either way occurs. Jump into a ice cold shower on a hot day sometime if you don't think it affects the fish. What we are really concerned about is the adjustment of a few degrees around room temperature when making changes in an aquarium. If the bag has sat in the area of the tank it is going to be introduced, chances are good the temperature is already very close.

However, temperature comes into play when the discussion changes to maintaining and breeding many species of killifish. *Nothobranchius* species of East Africa undergo large fluctuations in temperature from day to night. Those temperature extremes may actually have an effect on the incubation of eggs (discussed in further detail under breeding killifish) during the dry seasons as well as on the growth rate and breeding of the fish.

In West Africa, many of the rainforest areas are cooler than one would expect and those fish prefer to breed at temperatures in the mid-60s (15-20ºC). Yet in some areas of the deserts of the North American Southwest, the water temperature rarely gets below 95ºF (35ºC). Once again success with specific species of killifish requires a little extra research and knowledge about their native habitats.

During winter in Northeastern U.S. the water temperature drops to 40 degrees near the bottom and freezes on the surface. A number of species of killifish wait in the cold waters until Spring temperatures melt the ice and then their spawning mechanism takes over. In Argentina, members of *Austrolebias* and *Cynolebias* often live under a thin sheet of ice. Recent studies have shown these fish actually prefer cool water temperatures.

With all of this in mind, the hobbyist should worry much less about water temperature when fish are kept at room temperature in the home aquaria. The purchase of heaters and systems to heat a fish room should take into account the costs of electricity and/or the system used to heat the rooms. If a constant room temperature between 65ºF (18ºC) and 80ºF (26ºC) is maintained in the house, then there should never be a problem.

Rain Water

There is no doubt that rain water is a useful tool in the killifish hobby. Perhaps its main affect is on breeding killies where the addition of this water can initiate spawning behavior.

But rain water can also be a major problem. Air pollution can create unwanted organics to affect the water. For those who deal with acid rain in many parts of the world have seen whole ecosystems destroyed. This can also happen in the home aquarium when rain water is used.

In addition, rain water should never be used if it drains off most roofs, but should be caught free falling. Roofs catch a great deal of pollution, organic and inorganic that becomes concentrated in the runoff and will affect the purity of rain water.

Pure rain water has zero degrees of hardness and should be a neutral pH, but unfortunately most industries pollute the air and when water forms into droplets, that pollution becomes trapped and absorbed.

When a hobbyist chooses to use rain water, it really needs to undergo a filtration process with the use of charcoal. Charcoal will remove many of the pollutants from the water. It would not hurt to use a system that can hold peat

A distilling unit can be used in place of an RO Filter. Distilled water comes out with a neutral pH and zero degrees of hardness. It is considerably cheaper to purchase and produces about one gallon of water every 8 hours.

moss also. A canister filter with both types of media works very well, but should be recharged before each use to avoid any carry over of pollutants.

RO Water

There are a number of commercial companies that sell Reverse Osmosis (RO) systems to soften local water and water from wells to make it more palatable and to minimize mineral build-ups in the home. These same systems can be quite useful for hobbyists who live in areas where hard water is a problem.

Essentially, RO units use a semipermeable membrane that does not allow larger molecules to pass through. There are actually a number of steps the process entails including activate carbon to remove chlorine and inorganic chemicals as well as two separate filters to trap sediments.

For aquarium keepers, RO water has become an important piece of the fish care. Often it is a replacement for rain water because it does not have any pollutants. Hobbyists mix a percentage of RO water with tap water to create a specific hardness and pH to maintain the fish. Actually this is necessary since pure RO water would kill any fish by burning the membranes in the gills.

RO units work on pressure and cost nothing to operate. Units can be purchased in a variety of sizes that produce anywhere from one gallon a day to many gallons.

A simple system would be to connect a unit to a water pipe and allow it to drip into a reservoir, a large plastic container, trash can or barrel. That water is then mixed as necessary with tap water with a dechlorinator added.

An alternative to the purchase of an RO unit would be to obtain a water distiller. Essentially a distiller processes completely neutral water that has zer0 hardness. Once again, it must be mixed with water that contains minerals to reach the preferred chemical consistency and to be safe for fish.

Filtration: What Filters Work with Killies

Is it necessary to use filters in killifish tanks? Perhaps not, but it does make a real difference in water quality and that is what should the primary consideration.

There are two reasons to use filtration in fish tanks. The first is mechanical, the removal of particulate matter from the water. The second and more important, is to provide a place for the development of bacteria necessary in the conversion of ammonia and nitrites (* See the section on Cycling the Aquarium). There is no single 'best' filtration system. All have pluses and minuses. Most important is what a hobbyist feels comfortable using.

The following describes a number of filters to consider.

The Box Filter

Standard box filter using gravel and filter floss

KN PRESS - T.R. GRADY

There are numerous versions of the box filter ranging from home-made to elaborate commercial grade systems. The most common is a simple plastic box in several sizes regularly used by most hobbyists at one time or another. All work essentially the same. They use an inside platform with two tubes extending out of the filter, one for incoming air and the other to push water out of the box. The idea is to draw water through the filter media, pass it to the bottom of the container and then out the outflow tube. The filter media becomes a bed for nitrification bacteria and assists with the biological end of the system. Filter floss (polyester fibers), charcoal, lava rock and/or aquarium gravel (or another weight) are to hold the filter down and placed on the platform. A top keeps everything in place. Air tubing is attached to the intake tube and the simple system is in place.

The use of Lava Rock as a porous media is certainly a valid option. Lava Rock can generally be purchased in many garden and hardware stores as well as places such as Walmart and Target for use in barbecues. The rock itself will initially float until it becomes soaked and completely devoid of air in the pores. Bacteria will form in the Lava Rock and create an excellent environment for biological filtration. However, it does not make for a good mechanical filter. For that a pad of filter floss can be placed beneath the rock. The negative effects of Lava Rock will discussed in the section about the Debruyn Filter.

No matter what choices of media are used in box filters, one important factor in maintaining a healthy environment is to make certain it is cleaned regularly. This does not mean throwing out the baby with the wash water however. Because biological filtration is so important, maintaining the bacteria bed is necessary. Instead of breaking down the filter to its basic components and replacing them, the aquarist should carefully remove the filter floss (and lava rock if used) and separate it into a small container of water from the original tank along with the weight material (gravel or charcoal).

At that point, the plastic filter can be washed thoroughly, algae removed from the top and sides and the inside rinsed. If the filter floss is so caked in detritus, it is better to squeeze it in original tank water until it appears clear of the 'gunk'. The filter is then reassembled and the media placed back inside. At this point it would not hurt to add a little additional filter floss to the mix to give the bacteria additional space to expand.

Too often new aquarists want to change the filter floss when it gets dirty. This is a serious mistake. Every time the filter media is replaced or washed with tap water, the bacteria is destroyed and the biological cycle reset. This seriously endangers the inhabitants.

Sponge Filters

Sponge filters, a system that uses a variety of different foam-like materials are very popular with hobbyists. The filter itself is primarily a biological system and draws particulate matter to it. It is still one of the best filters available, particularly for small tanks. Filters are available in various sizes that work well for everything from 2-1/2 to 20-gallon tanks. For larger tanks the hobbyist can use multiple filters to accomplish the job.

Over time sponge filters can become clogged and the only real way to clean them is to remove them from the tank and rinse them out. It is highly recommended that this be done in the following manner:

Fill a small bucket with water from the tank. Separate the sponge from its base and immerse it in the tank water. Using hands, perhaps covered in gloves, strip any algae or heavy detritus from the outside of the foam.

Follow this by squeezing several times to move water through the sponge. Obviously this will create a huge cloud of murky water as the detritus is removed. Squeeze it one final time, but do not release the sponge until it is removed from the bucket. This allows the foam to retain much of the bacteria that has become established.

If the sponge is rinsed out under tap water, there is a very good chance the majority of the bacteria will be lost. There are some who suggest tossing the foam into a washing machine to clean it. This is a huge mistake. Sure it comes out clean, but also kills all the useful bacteria and the biological system has to begin again.

Lava Rock Drip Filters

In the 1990s, Henri DeBruyn of Belgium came up with a fascinating filtration system that works exceptionally well for killifish, in particular for some of the rainforest species of *Aphyosemion* and *Rivulus* that are a bit more difficult to breed. Unfortunately each filter must be built by hand and it is not currently available commercially.

One of the primary advantages of this system is it allows for a higher concentration of oxygen to enter the system via the drip method. The oxygen enhances fish health. The originator of this system observed in nature that the water surface versus the number of fish in the natural habitat was large and water was highly oxygenated. In essence this system creates a constant breaking of the surface tension by dripping and allows for the exchange of gasses to be increased. In some ways it could be considered like the dripping of rainfall from the canopy above.

The "DeBruyn Filter' uses lava rock as the primary biological filtration media, but the system acts like a wet-dry drip filter. The basic assembly, made of vinyl gutter, sits directly on top of the aquarium. Water is pumped from the aquarium through an uplift tube via a simple air hose. The water spills out over the

Ken Normandin of Florida has written a superb article on creating this filter along with all the advantages and disadvantages. It can be found on the American Killifish Association website at http://www.aka.org/UserFiles/File/debruyn_filter.pdf

lava rock and drips back into the tank through a series of holes drilled into the base.

Like all filters there are benefits as well as detriments to this system. The positives may outweigh the negatives, but that is a decision for each hobbyist if they decide to try the system.

KN PRESS - T.R. GRADY

Advantages:

- Removes biological waste from the water
- Does not impact the hardness of water
- Oxygenates the water
- Minimal maintenance once bacteria is established
- Overfeeding is not an issue

Disadvantages:

- High rate of evaporation
- Bacterial muck gathers on the bottom of the aquarium over time and needs removal. (This muck is inert, but unsightly)
- Does not remove the uric acids from fish wastes.
- The pH of a tank lowers over time.
- oIf the system is shut down for more than 2-3 days, water quality degenerates and requires water changes to re-establish the balance.

Power Filters

There are a number of options running from 'behind the tank' filters to larger self-contained systems or wet-dry filters. All of these filters use a motor of some sort to power the flow of water. One of the real benefits of these filters is the wide variety or combinations of filter media that can be used. All of them use a form

of biological filtration, but most also handle mechanical as well. It's simply a matter of choosing the media an aquarist prefers.

A basic 'behind the tank' filter can use pre-made cartridges with charcoal contained in a fine particle polyester bag and frame that usually is covered in a floss-like material. The addition of lava rock into the filtration box adds an additional location for nitrification bacteria to form. These filters come in a variety of sizes and designs, some of which include additional elements to assist in oxygenating water and/or protein skimmers. Most use a magnetic 'motor' to move the water up a tube into the box.

External Filters

The hobbyist can choose a larger external self-contained system which has a number of chambers for filter media. Depending on the system, it can hold pre-made types of media or the hobbyist can add whatever media is preferred.

One of the benefits of this system is the ability to filter a number of small tanks at the same time if the aquarist designs an interconnected flow system. The benefit of this is a larger volume of water flows through every tank reducing the possibility of massive water quality changes. The negative issue is that disease can spread to several tanks at the same time if one tank becomes infected. The hobbyist has to weigh the good with the bad.

The ultimate version of external power filters would be a 'Pool Filter'. Yes, a swimming pool filter is an option. These filters can be used to create a flow-through system for a large number of drilled tanks and connected via a series of PVC pipes. The basic filteration is done via sand placed inside the canister. Often thie system uses an ultraviolet light to kill bacteria.

There is a huge turnover of water volume and it can be designed to add a continuous flow of 'new' water into the system as older water is drained into the sewer system.

One issue is how to eliminate chemicals from the direct infusion from the local water system. One option is to use a 'drinking water filter' which can remove most unwanted chemicals in-line before it reaches the pool filter. To replace the lost minerals wanted in the system, the water can run through a reservoir contained peat moss, calcite chips or whatever is desired, Admittedly, this requires some plumbing skill to create. A system of this nature will rarely have problems with ammonia and nitrates once it is established because of the turnover of fresh water replacing the waste water.

Aeration & Air Pumps

For nearly every living thing on Planet Earth, oxygen is the most important element for survival. It is no less important for tropical fish, just the biological method applied is different. The hobbyist needs to create an oxygen enriched environment for the captive population to maintain the best possible health of the fish. There are a couple of factors that need to be understood in order to accomplish this.

- It is important to understand the percentage of oxygen in water vs. breathing air. The air humans breathe contains slightly more than 20% oxygen, while water only contains about 1% of the life sustaining gas.
- New hobbyists often misunderstand the way oxygen is added to water and think that bubbles from an air pump is what injects oxygen into the water. In reality, the primary gas exchange happens at the surface of the water, not by bubbles rising through it. The larger the surface area, the more the exchange can occur.

- The gas exchange is primarily carbon dioxide and oxygen.
- Surface tension is another consideration. Stagnant water can build a film and inhibits the continuous exchange of gases. Instead stagnant water can lead to very poor oxygen levels in the water.
- Water movement is the single-most important component necessary for the highest concentration of oxygen. This can be accomplished in any number of ways. The stream of air bubbles through the water column breaks the surface tension and creates a wave action which actually expands the surface as well as breaks up any scum which can form and block the gas exchange. Alternatively, external flow of water cascading from outside filters accomplishes the same result. The DeBrutn Filter wimulates rain drops to also allow for better oxygenation.
- Plants also assist in oxygenation. Plants need carbon dioxide to survive. As they absorb the CO_2, plants release O_2 into the water. The benefit comes from the fish who give off CO_2 as part of of their natural respiratory process and take O_2 from the water. (Plants and a variety of uses are discussed in its own section).

How Oxygen enters the system.

As already discussed briefly, a flow of air bubbles through a variety of box and foam filters moves water throughout the entire system and creates a circulation in which water is forced to the surface and then downward. The gas exchange occurs along the surface with the movement of the water. This is the primary reason the use of aeration devices is important.

These can range from small diaphragm air pumps to elaborate linear units and fan-based blowers. Much of the hobbyist's choice has to be based on a combination of the size and number of tanks in a fish room along with the cost of operation. For someone with only a few tanks, a larger diaphragm pump may be the best answer, but as the number of tanks expands, then alternatives must be considered for financial reasons if nothing else.

• **The DeBryun Filter** (discussed in the section on filters) may be the best system for oxygenation. Water flows over lava rock in a wet-dry system and is oxygenated by this movement. In addition, as the water drops back into the tank it acts a little like rainfall and constantly breaks the surface tension and creates water movement where the gas exchange continues.

• **Behind the tank** filters also oxygenate the water on two levels. First, rapid movement of water into the filter media creates additional surface space for the exchange. In addition as the water spills back into the tank it breaks the surface tension and creates wave action. A few of these filters have a protein skimmer which is essentially a fine mist of air bubbles inside the filter box. A small amount of additional O2 is added via this skimmer.

• A few of these filters may have a **water wheel** as part of the design. This can also add gas exchange surfaces.

Heating Killifish Tanks

As discussed in the opening of Care of Killifish, there is a wide diversity of temperature variation in the natural environment for the well over 1,100 species of killifish. Temperatures range from extreme heat in the American desert Southwest to freezing cold in Argentina. However the vast majority of killies do well at what is considered normal room temperature of 65ºF to 80ºF (18ºC to 26ºC). If specific needs are necessary for success with certain groups, that information is contained under the description of such.

However, a couple of things should always be considered for almost all tropical fish when it comes to heating individual tanks. It is not uncommon for the small, inexpensive tank heaters to go bad and cook a fish tank. There are several variations of in-tank heaters. Some hang from the sides while others are submersible. The same heaters may just not work or will have uncontrolled variations in the temperature produced. With this in mind the hobbyist should consider what type of heating to use.

Another option that can be considered is a heating pad system or heating cables. Pads are installed beneath the tank while cables can be placed beneath the bottom substrate. Neither are commonly used in the killifish hobby.

In heating an entire area, a couple of interesting things should be kept in mind. Heat rises, so the ceiling will be much warmer than the floor. For some breeders, this is a valuable tool, keeping cool water killies on lower racks and those desiring warmer temperatures up high. If the aquarist wishes for the room to be held at a more stable temperature, ceiling fans can be installed.

Lighting and Illumination

In nature, the sun provides intense lighting of a full spectrum of colors even when partially occluded by clouds. Light is necessary for the growth of life on all levels from the smallest creatures to vegetation which feeds larger creaturese. Algae creates the largest percentage of oxygen on Planet Earth and without sunlight, there is no algae. In terms of killifish, the intensity of the lighting is far more important for maintaining plants.

Killifish prefer moderate illumination with a few exceptions such as desert pupfish which are exposed to strong direct sunlight during the day. Just recognizing the vast majority of killifish species come from tropical or semi-tropical regions with heavy terrestrial and aquatic plant growth demonstrates the fish prefer to remain in dimmer places of the environment. Most killies can be considered as prey for larger predators and hiding in dense vegetation is a defense mechanism.

Artificial lighting for killies should operate from ten to fourteen hours a day. The day/night cycle can easily be maintained with inexpensive timers available almost anywhere light bulbs can be purchased. In nature, most plants are found in shallow whater and are subject to the full spectrum of light. This type of light equals Kelvin scale ranges from 5500K Red) to 7500K (white). Intensity is actually based on the wattage used by the bulbs. The hobbyist should seek out full spectrum lighting for the best results. This is usually documented on the packaging of the bulbs.

Why is light so important?

The obvious reason is so the fish can be seen, but that is not by any account the most important reason.

Plants use carbon dioxide in conjunction with water while under light to create sugar used to feed the cells. This is called photosynthesis. Oxygen is released during the process. Alternatively, plants also need to breathe. During the 'Respiration' process, which is when energy is released from the sugars, carbon dioxide and water is produced and oxygen taken in. Respiration takes place all of the time. Since photosynthesis can only take place with the presence of light, a cycle of light and dark is important to the process. The complete cycle maintains a balance of CO_2 and O_2 in the environment.

In the home environment there are a variety

Aphyosemion elberti Ndouzum

Photo: Richard Pierce

KN PRESS - T.R. GRADY

of options for lighting ranging from overhead shop lights to elaborate LED systems. Perhaps the most common lighting used by hobbyists is florescent lights directly above the tanks. These light can promote plant growth under optimum conditions and heavy algae growth under less than ideal. Recently the use of LED lighting has gained considerable support and there are a number of aquarists who swear by it.

Florescent Lighting:

This is the most common form of lighting for fish tanks no matter the genus or grouping. Commercial companies produce a variety of aquarium covers which use replaceable florescent bulbs and the aquarist has a variety of options on how to use these lights. It is not uncommon to find fish rooms where every tank has its own cover, but the cost in electricity can be high.

Alternatively many hobbyists prefer to use overhead shop lights which use 3-4 foot long bulbs. Some of these can be placed directly on top of the tanks and cover several, while the same lighting systems may also be hung from the ceiling to illuminate a wider area. How much light the aquarist wants directed to the aquaria is key to which option is chosen.

Plant lights sold in retail stores have a tendency of promoting algae growth and should be avoided. Most large box stores carry a variety of bulbs from 25 watts to 40 watts that can be used and are relatively inexpensive. The color spectrum should be white (11) and warm (32) tones.

LED Lighting

LED light strips can be expensive to purchase, but they last a long time and use minimal electricity which reduces the overall cost of using this method. For optimum plant growth 6500 Kelvin is considered 'Daylight' and tends to have the necessary mix of red and white light. A variety of companies produce LED lights that can be modified to create homemade covers. Commercial aquarium LED lighting systems are expensive.

The LED lights have been touted as one of the best options for plant growth.
Home-made LED Lighting System:

A homemade LED system that can be used over a number of tanks can be created at a fairly reasonable cost.

Community tank   Photo: Richard Pierce

Vegetation: Green Matters

The Value of Plants

Breeders in many cases tend to keep killifish in 'bare' tanks, aquariums with mops and filters and little else to simplify the care. This is not the natural habitat for these fish, in fact most live in dense vegetation in the wild.

Killies need the plants and roots as places to hide from predators and to leave their eggs. Nothobranchius males tend to set up near the stalks of rooted plants that reach to and through the surface. The females then approach the males when ready to spawn.

Aphyosemion species of West Africa remain close to the banks of streams and brooks where vegetation is the most dense. This is true around the world for most species unless they are in isolated locations where natural fish predators are rarely a problem (the Devil's Hole Pupfish, *Cyprinodon diabolis*, would be an example).

There are a number of popular plants that are valuable for killie-keepers. For killies a real mix of floating and rooted plants are worth examining and most only require a minimum of care.

Plants are a useful part of a balanced system, releasing oxygen into the water and using the carbon dioxide from the fish' respiration. In general, plants provide a sense of security for all fish.

Community tank   Photo: Richard Pierce

Floating Plants

Hornwort: A popilar and common, useful floating plant that can easily overrun a tank is *Ceratophyllum demersom*, the well known Hornwort.

This plant sends off branches and the older segments will fall away and form a se[srate plant. This is an excellent plant for fry tanks, but certainly adult fish love to hide in it and as a natural spawning media little is better.

Hornwort is a popular plant for many killifish species to hide and for you fry to escape predators.

Water Sprite: *Ceratopteris thalichtroides* is one of the most popular floating plants used in the hobby, but it can also be frustrating for some. It is sensitive to water conditions and needs a strong source of light. However the benefits for killies may outweigh the occasional problems with keeping it. It is a great plant for many species to use as a breeding location and the fry and juvenile fish can use it for security.

Swater Sprite   **Photo: Richard Pierce**

Susswassertang: *Lomariopsis cf. lineata -* **This fern is** an excellent choice for fry to use as a hiding place, Susswassertang is a water column feeder, meaning it does not need to root, but instead uses light as it's primary source along with nutrients in the water. It can be tied to driftwood and will attach over time and become a lovely decorative plant.

Susswassertang **Photo: Richard Pierce**

Java Moss: Whether this should be considered a floating plant or not, Java Moss, *Vesicularia dubyana*, is one of the most essential plants found in killifish tanks. Java Moss forms large networks of tiny runners that become as thick or sparse as the hobbyist needs. It is used as breeding vegetation, in fry raising tanks and as a decorative part of community tanks. Wrapped around driftwood and rocks, it forms a blanket that becomes a lovely green carpet. Newly hatched fry trays should contain some Java Moss if for the microorganisms that live within.

Java Moss   **Photo: Tom Grady**

Duckweed: Probably the most despised plant in the hobby, Duckweed (*Lemnaceae minor*) can be extremely difficult to eradicate once it becomes established in the fish room.

All it takes is a single plant to reproduce, which it does by splitting. If a single piece of duckweed gets into a tank, chances are very good the entire surface will be covered in a thick carpet in a few months. It is easily transported from tank to tank by nearly anything that touches it. Several fish eat duckweed. Goldfish, some plecos and a number cichlids will eliminate it from a tank.

It does provide a couple of real benefits for the aquarium. First it is a good indicator of water quality. If the plants begin to turn white and die, then there is a real problem. Secondly, it is a viable water purifier, removing excess nutrients and helps in the prevention of algae.

Plenty of microorganisms find a home among the plants.

Duckweed becomes a real problem if left to its own and will carpet the entire surface of the tank up to a depth of an inch if given the opportunity. It is best to simply not allow it to become established in the home aquairum unless it is desired.

Pond Weed: *Najas guadelupensis* is another floating plant that is fairly easy to cultivate and provides excellent cover for killies, both fry and adult. It prefers bright light and has little difficulty with most water conditions, accepting lightly brackish to fresh.

Najas
Photo: Richard Pierce

Plants with Rhizomes

A rhizome is a modified stem of a plant usually found underground, but In aquariums is commonly seen running along the bottom of an aquarium. This plant often send out roots and shoots . The rhizome also retains the ability to allow new shoots to grow upwards.

If a rhizome is separated into pieces, each piece may run into a new plant. It uses the rhizome to store starches, proteins, and other nutrients. These nutrients are useful for the plant when new shoots must be formed or when the plant dies back for the winter

KN PRESS - T.R. GRADY

Java Fern: *Microsorum pteropus* is a popular plant that can be planted in the gravel or attached to driftwood to provide a nice decorative background for any tank. They thrive in a variety of conditions and require only moderate light. They normally reproduce by forming leafs leafs that drop off to start new plants, but also by rhizome division.

Anubias: There are numerous species of *Anubias* which can be part of any ecosystem.

These slow growing plants are excellent for the floor of most community tanks, but also do well when planted in a small pot of some sort. Very few fish eat the tough leaves which sprout at best about every three weeks.

Anubias do well when grown above the surface of the water as well as submerged. There are many species of *Anubias* available in the hobby

Anubias nana   Photo: Richard Pierce

Rooted Plants: Obviously rooted plants can be planted in any gravel or sand. However, for aquarists who prefer not to use the substrates, a simple alternative using baby food jars or any number of other inexpensive containers works quite nicely. Simply add some peat moss to the bottom of the container and cover it with sand. Plant your plant.

Crypts: Members of the *Cryptocorynes* are another beautiful addition to community tanks and are a long leafed group of plants. Most crypts come from tropical waters in Asia and New Guinea and are found in slow moving streams and rivers, as long along the banks of lowland forest pools. *Cryptocoryne wendtii* is one of the most popular varieties.

Many species of Crypts are reproduced submerged in the aquarium, but in Oriental nurseries it is common practice to emersed in dampe soil. Emersed plants have a tendency to lose their leaves when planted in a tank. Nearly all species of Crypts can be grown this way.

Cryptocoryne wendtii   **Photo: Tom Grady**

Vallisneria: There are a few beautiful species of Tape Grass or Eel Grass are available on a regular basis in pet stores. Most do well once planted in the substrate and will spread by runners. On rare occasions female plants will sprout delicate flowers on a long stalk. They are also capable of producing fruit, small capsules that contain seeds. The largest forms can reach a couple of feet in length and smaller ones 10-12 inches.

The Corkscrew Val, *Vallisneria spiralis,* is very popular because of its appearance.

Vallisneria   Photo: Richard Pierce

Sword Plants: Amazon Swords and Brazilian Sword Plants are two of the more popular species of *Echinodorus* regularly found in community aquariums, but not the only ones. These plants can become huge under the right conditions, with dozens of thick broad leaves stretching to the surface.

Bright lighting and injected CO_2 will cause these plants to grow at a very rapid pace. They require nutrients and fertilization is important.

There are numerous variations of swords available including the Ruffled, Narrow Leafed and Red Rubin.

Photo: Richard Pierce

Hair Grass: This dwarf species is very useful for dmall killies as a place to escape predators. It is a bottom covering plant that works well in filling out the foreground and lowest regions of the tank.

It propagates by sending out numerous runners that branch from the root area. It grow quite fast. This is a sunmersed plant that will grow its leaves above the surface in low water tanks, but it rare gets taller than gour inches. It does well in a neutral pH and prefers temperatures in the mid 70s Fahrenheit. Water hardness of 4-8 KH is best. While not essential, it does best when fertilized.

It is also used by many of the switch spawners as a place to leave their eggs.

Feeding Killifish: Live, Frozen or Flake?

Any discussion about feeding killifish usually begins with someone asking if they only eat live foods. The simple answer is no, but the best answer is much more complicated. Killifish definitely prefer live foods, but that does not mean they must be fed exclusively that way. In fact most species will adapt and soon accept anything including dried flake foods and freeze fried options. The question then becomes what is best for the fish.

In the wild, killifish are primarily carnivorous or at worst omnivorous, but there are no true vegetarians except *maybe* a few Pupfish, but even they will happily take live offerings. The food most killies seek out is a combination of insects, protozoans and small fish. Ants, mosquitos, worms and insect larvae compose the majority of foods accepted by killies in the aquaria.

However, there are a few large predators which must be viewed in a different way. *Nothobranchius occellatus* in East Africa and *Megalebias wolterstorffi* in South America are good examples of true predatory killies and both enjoy a good repast on other killifish or whatever else can fit into their mouths. In the case of *N. occellatus*, keeping individuals, even males from females, separate is a good choice. They will kill and eat anything in their tank unless they are breeding. Sometimes it does not hurt to maintain a tank of 'feeder guppies' to use for these predators.

Without a doubt emulating nature is almost always the best choice. Just the shape of the vast majority of killies indicates they are surface feeders for the most part. Mouths are upturned, eyes are either on the upper sides of the head or near the top. These are environmental adaptations that allow the fish to seek food at or near the surface of the water. Many species are insectivores, seeking ants and other insects that fall to the surface. Strangely, it is rare killies will eat ants in captivity despite analysis of stomach contents of wild fish reveals ants are eaten. Nearly every species feeds on a wide variety of water based animals, copepods and larvae. And of course there are some true predators who feed on other species of fish ranging from killies to cichlids.

An important factor in breeding and raising quality fish, whether killies or otherwise, is a varied diet. The caretaker needs to recognize restricted feeding of one or two types of food is not healthy for long term success with any species. There are dietary needs all fish require for their optimum health and growth. While the most popular dry foods are well infused with vitamins and other dietary components, over time the same foods become less desirable because while they may keep the fish alive, the offerings can lead to certain deficiencies not yet understood or recognized.

Another consideration for the use of live foods is fish behavior. In the wild killies are constantly hunting for food and feed when the opportunity occurs. If we only use dry food, the fish may lose that activity and this could lead to 'spoiled' killies - one in which the animals only become active when food is added to the tank. If we really consider it, the two major activities of fish in nature are feeding and reproduction. Those feeding activities provide the fish a reason to 'exercise'. How much of this is really necessary for killifish health is speculative, but if we want to emulate nature, we need to do more than add plants to an aquarium, we should examine as many factors as possible.

This section will examine some of the easier to cultivate live foods for both adult fish as well as fry. In some cases multiple techniques are suggest to provide alternatives. The method of cultivation is dependent on what space and euipment is available.

Frozen: Almost all killie keepers maintain a good stock of frozen brine shrimp and possibly blood worms on hand and most feed these foods to their fish on a daily basis. *Daphnia* and baby brine shrimp are also available in frozen form. Techniques vary from simply cutting off frozen chunks and releasing them directly into the tanks to allowing the frozen food to thaw in a container of water and spritzing it into fish tanks. Either way works fine.

Flake Foods: Killies will eat flake foods and do so as well as any other group of fish if they are properly introduced to it initially. A little starvation never hurts to get them feeding this way. Go a few days without any other food in the tank and it is pretty much guaranteed the killies will accept the flakes readily from that point onward.

Freeze-dried Foods: While killies will sooner or later accept the freeze-dried offerings, they are more reticent about the feedings. In a sense, does beef jerky really sound like a mouth watering meal?

Cyclop-Eeze: These are interesting biologically engineered organisms from the copepod family which have high concentrations of omega-3 fatty acids. As an additional option for the foods taken by killies, this product is designed more for fry and juvenile fish than adults. Many hobbyists swear by Cyclop-Eeze and for that reason it is included in this section.

The origination of the food comes from arctic waters in northern Canada, but is now cultured in high salinity ponds by Argent Chemicals Laboratories. It comes in both a freeze-dried form and a frozen one. The size of the particles is very small and apparently fry will take them. The producer of this product claims it enhances reds and blues in adult fish because of the concentration of Highly Unsaturated Fatty Acids (HUFAs).

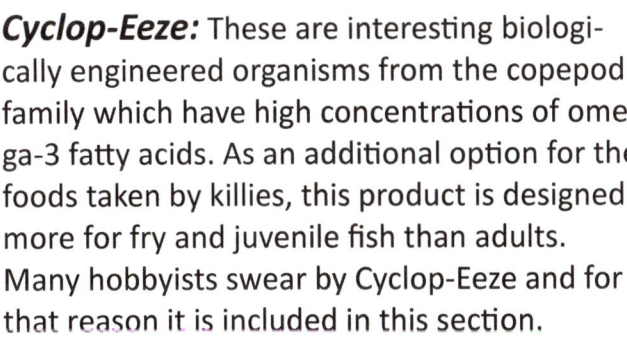

- 39 -

Culturing Live Foods

Brine Shrimp

Most likely the most commonly used food for killies, Artemia is available in several forms ranging from frozen to freeze dried to live. Frozen and dried foods are discussed in the previous section, but live brine shrimp is a staple of the hobby. Culturing brine shrimp and growing them to an adult size is possible. But realistically, most good aquarium stores carry a supply and the adult shrimp be purchased at a relatively inexpensive price.

Can brine shrimp be raised to adult? Yes, but it generally does not provide enough shrimp to be a regular feeding unless large pools of salt water are maintained and some form of nourishment is provided for the shrimp in order to grow. Generally, financially this is not worth the effort, but it can be accomplished in plastic containers set up outdoors. The water needs to as salty or more so than the ocean.

There are commercial foods that can be fed to the shrimp, but the development of algae and rotifers is really the best option. Actually, the brine shrimp will feed constantly on algae and become tinted green as they grow. It just requires a great effort to mass produce this live offering.

Baby Brine Shrimp: For this discussion, we are talking about culturing brine shrimp, primarily as a supplemental food or more importantly the primary food for baby killies.

However baby brine shrimp can be offered to almost every size of killies except perhaps the largest.

Brine shrimp produce cysts in the wild which float at the top of a number of salt lakes such as Great Salt Lake in Utah as well as in the San Francisco Bay. Around the world are numerous other locations. The cysts are harvested by commercial companies, packaged and sold in various quantities from one ounce to 15 ounces. Larger quantities are available. These companies collect the floating rafts of eggs which are then dried out and finally enter a cooling period. In nature brine shrimp do reproduce both through the cysts and also live birth. This is a mechanism developed to ensure the survival of the species. When the salt levels exceed 150 ppt (or four times the salinity of the ocean), the shrimp release the nauplii live.

> **Note:** Brine shrimp eggs can be stored for very long periods (years) by freezing them.

As a hobbyist we want to hatch the eggs on a daily basis. Depending on the temperature of the area where the brine shrimp are to be cultured, the time it takes to hatch ranges from 24-48 hours. Cooler temperatures result in longer delays. Room temperature of between 70ºF and 75ºF will normally take from 24 to 36 hours to hatch. Higher temperatures will result in faster hatches.

There are any number of ways to hatch out BBS. It can be done in long low trays or soda bottles. Commercially created hatching containers are available. One of the simplest techniques simply uses gallon jugs with a strong air

flow with an air hose immersed to the bottom.

Salt water is prepared a variety of ways. Many types of salt from kosher to sea salts may be used to prepare the mixture. Aquarium sea salt can be expensive, but may actually produce a better percentage of nauplii. A salinity of 35-40 ppt (or a specific gravity of 1.024-1.028) is best.

Aeration is important. A high volume of air pumped through the container will create good circulation and keep the eggs and later the brine shrimp suspended in the water column. This will actually allow the nauplii to survive longer. If this does not happen, then the baby brine will lie on the bottom of the container and die from lack of oxygen and foul the water quickly.

To harvest the shrimp, whatever method is chosen becomes a basic siphoning of the shrimp through a brine shrimp net or cheesecloth (or any cloth fine enough for the water to pass through and leave the shrimp behind.) The strained shrimp should then be placed in a fresh water filled container essentially rinsing residual salt from them. This can be poured through the brine shrimp net again and then placed back into a different fresh water filled container. Feeding killies the nauplii can be done with a baster withdrawing water containing the shrimp and squirting the food into the tanks or trays.

De-Capsulized Baby Brine Shrimp: Baby
Brine Shrimp/Nauplii are the essential food for newly hatched killifish fry. On the market today is a product which can be a substitute under the right circumstances. Purchased direct from a number of commercial companies, the BBS are actually eggs which have had the cyst-capsule removed leaving only the shrimp itself. It comes in what appears to be a powdered form and can be sprinkled on the surface of the water in a fry tank. It is supposed to be equally nutritious to the live BBS.

Some hobbyists also use a technique to remove the shell of the brine shrimp cysts and feed the result to the young killies. Essentially it requires the use of chlorine bleach and a fresh water rinse. The decapsulated shrimp are then placed in a salt solution to 'hatch' in a similar method to simply hatching the cysts. These nauplii can survive the process, but even if they do not there is really no difference from feeding killie fry with live nauplii. The one real benefit is the residual shells no longer have to be cleaned from tanks.

At times in the past, brine shrimp eggs from Salt Lake or San Francisco Bay were not hatching well (at one point nearly a zero hatch rate). This method proved to be a very useful option.

It is very important to note, fish fed exclusively on *Artemia* will not receive all of the essential nutrition necessary to remain healthy throughout their lives and it has been reported that some of the results include lowered growth rates, decreased reproduction capability and juvenile weakness and death as they near maturity. For this reason, breeders may wish to add supplements to the newly hatched Artemia. Several commercial brine shrimp companies offer HUFA enhanced concentrates

Decapsulized Brine Shrimp will hatch and swim, but can be fed directly into the tank.

- 41 -

KN PRESS - T.R. GRADY

to add to the hatching process for BBS.

Immediately after hatching, the nauplii remain within the hatching membrane and absorb the remainder of the yolk. This stage takes a few hours before the BBS emerge. Artemia are filter feeders, but need to molt before they are able to begin feeding. This occurs around 12 hours (again at 77-80F) following the initial hatch. With this information in mind, it appears a feeding rotation of 48 hours is a good choice if using supplements to enhance the food value.

Daphnia & Moina

Daphnia is a generalized term in the aquarium hobby that depicts a group of small crustaceans that are relatively easy to culture as a regular live food offering for most species of smaller fish in general. Killifish take them eagerly. Two types of *Daphnia* are commonly used, *D. pulex* and *D. magna*. The *D. magna* are larger. Another crustacean that is in use is *Moina*, a similar and also easily cultured live food. *Moina* are smaller than *Daphnia*, about half the size of D. pulex. Both of these small crustaceans are cultured in the same way.

The life span for *Daphnia* is normally 5-6 months, but under colder conditions can reach as long as 13-14 months. In nature *Daphnia* generally feed on unicellular algae and the byproducts of detritus, primarily bacteria and protists (single celled animals and plants that do not form tissues.). The hobbyist needs to feed *Daphnia* with alternative options. Yeast is a viable choice and if a culture of algae is available, then that is an excellent supplement. Using the below technique, bacteria is also produced and will add to the overall nourishment.

Culturing *Daphnia*: Large (30-gallon/113.5L) heavy plastic bins are readily available at most large department stores and make excellent containers for a variety of hobby related needs, in this case, culture containers for

Daphnia. Fill this bin with water removed from fish tanks as part of water changes. There will be some detritus in the water and that is actually beneficial in getting the bacteria initiated. Water temperature can range from the low 50s (10-15C) Fahrenheit well into the 70s (20-25C) and Daphnia will reproduce.

Snails are essential in maintaining a balance within the culture system. Whether it is Rams-

Daphnia ready to be fed to a tank of killifish
Photo: Tom Grady

horn snails or pond snails, they perform a twofold function. First they consume the Daphnia that die and second, they produce wastes which feed the bacteria which in turn feed the Daphnia. This is a very nice cycle of life. While this cycle could never produce enough *Daphnia* to feed tanks of fish, it will allow the culture to exist when unattended for longer periods of time. It can actually exist at a low level for months. This is where the feedings of yeast and algae come into play.

Because *Daphnia* are considered filter feeders, the use of dry active yeast (Fleischman's and Red Star are two commercial brands) is fine to provide a source of nutrition. Using warm (not hot) tap water in a small container, enough yeast should be added to cover the surface and then stirred into a milky consistency. A small amount of this mixture is poured into each of the daphnia cultures only when the previous addition has cleared, generally in two to three days.

It is an absolute necessity that *Daphnia*

cultures are harvested regularly. If they are not harvested, the culture will crash and a new one will need to be started from scratch. Unfortunately, *Daphnia* cultures crash occasionally for no apparent reason even though they may have survived for very long periods of time, even years. It's possible some external environmental change may cause the die-off. Sometimes when this occurs, all of the cultures seem to be affected at the same time and the entire food source disappears. The other thing to avoid is the use of some ammonia locking products. These will kill the *Daphnia* and must be avoided at all costs.

Harvesting and feeding *Daphnia* couldn't be

Rubber bins of this size (30 Gallon) make for excellent containers for Daphnia, Moina and also as a place to raise earth worms and nightcrawlers. They can also be used as a larger rearing tank for most species of killifish.

simpler. Simply swirl a fine mesh net gently enough not to stir up the detritus on the bottom in as many of the containers as needed. The food source can be rinsed under flowing water and then released into a water filled feeding container. A turkey baster can be used to withdraw the *Daphnia* and squirted into each tank.

Adding *Daphnia* to new-born killifish fry tanks can be valuable. The filter feeding of the

adults helps clean the water. As the adults shed their exoskeleton, live baby *Daphnia* (nauplii) are released and are small enough for most killie fry to eat.

Because *Daphnia* is a crustacean like brine shrimp, there is some concern that an exclusive diet will cause some digestive problems, perhaps blockages, it is best not to overfeed any combination shelled foods. As part of a well balanced diet, *Daphnia* can be considered an excellent source of nutrition.

Mosquito Larvae

No one wants to be bitten by mosquitoes, particularly knowing the variety of diseases these winged vampires carry including Nile fever and Yellow Fever. However many species of killifish love a good feeding of mosquito larvae and it will help initiate spawning. This is not a food that can be cultivated in a fish room. Instead it must be collected at some small risk of providing a meal for the insect.

Mosquitoes will lay their eggs in just about any pool or puddle of standing or stagnant water. The larvae swim to the surface, but will dive to the bottom when they perceive any danger. The idea is to look for the squiggling larvae and then harvest them.

If a hobbyist wants to simplify their collection of mosquito larvae, then a large pool of water (anything from a 30-gallon plastic tub to a small swimming pool) should be placed in a shaded area fairly far away from any houses. The use of some decomposing matter such as grass will draw the mosquitoes to the container. To reproduce the mosquitoes require blood, human or otherwise. An open invitation to invite them inside is not wise. Alternatively, the hobbyist may consider seeking out natural stagnant water. In the Spring, mosquito larvae can be found in the thousands when the eggs hatch.

To harvest the larvae, it is a simple process to swirl a fine mesh net in the container to collect them. The hobbyist might consider attaching the net to a long pole to reach further into stagnant pools.

There will be a number of sizes from very small larvae to 'bullheads' - larvae very close to the stage where they become mosquitoes. It is important to understand that other insects will lay their eggs in the same water and it's necessary to be cautious about what is introduced into the aquariums. Dragonfly larvae will feed on fry and even larger juvenile fish for example. It is possible that *Daphnia* might also be found in the same waters. Separating the *Daphnia* to use as a culture starter works quite well.

The use of a turkey baster to remove the larvae is recommended. While it is a little cumbersome, it is the best way to be certain only mosquito larvae is fed to the fish. The larvae can be squirted into another fine mesh net and then rinsed under slow flowing tap water before adding them to a larger container filled with clean water. The turkey baster can then be used to spray the food into tanks.

For many species of killifish, mosquito larvae are a superb food for bringing them into breeding condition. Nothos and the larger *Fundulopanchax* will not just eat, but attack this food.

The Worms

Killifish hobbyists are probably the primary group of enthusiasts who maintain numerous cultures of worms to feed their fish. Worms are designated as everything from Vinegar eels for the smallest fry to red wrigglers for large fish like the Blue Gularis. All are cultured in fish rooms or other areas of the house to the

The purchase of an old refrigerator at a yard or garage sale will help maintain relative peace in the household. Place it well away from where food is kept. The temperature can be adjusted to whatever is necessary to keep the cultures.

distress of spouses. Refrigerators often become home to a number of live food cultures such as black Worms, microworms or tubifex, not to mention frozen foods in the freezer.

White Worms

White worms, *Enchytraeus albidus*, are among the most popular, yet often most controversial live foods hobbyists feed to their fish. One of the major questions often raised is whether or not white worms are too fatty and do not promote healthy maintenance of killies, particularly if used exclusively. The simple answer for that is - do not use white worms as the only food. Feeding them as part of a varied diet is a viable and actually very good choice. The true value of white worms is in conditioning killies for breeding. Along with black worms and mosquito larvae there is no better conditioning option.

There are two forms of white worms in common use in the hobby - the commonly known 'white worm' and it's smaller cousin the Grindal worm, *E. buchholzi.* A number of species of *Enchytraeus* exist and any or all may be in use in the hobby, but the above two forms appear to be the most common. There may be as many ways of culturing these worms as there are hobbyists, but most techniques are based on a few simple concepts. Provide a media for the worms to become established and feed them a steady diet. Many forms of *Enchytraeus* lack sexual organs and the primary way of reproduction is by fragmenting, simply splitting off segments of their bodies.

All forms of white worms can be cultured the same way, but hobbyists always look to minimize the work necessary to prepare the actual feeding. Below are two methods used to propagate cultures large enough to be used on a regular basis.

The Soil Method works fine for larger species of White Worms, while the Foam Box Method is excellent for the smaller worms such as Grindals and is described under the Grindal Worm section. It can also be used for white worms, but the container must be on a larger scale. Anything from a shoe box to a styrofoam box works.

Soil Method: One of the most popular ways to culture white worms is the soil method. The soil used can be garden forms of peat moss

sphagnum exclusively or up to a 50/50 mixture of a rich garden soil (African Violet Soil is a good choice) mixed with the peat moss. The mixture needs to be quite damp, almost wet. A good rule of thumb is if the water puddles or the soil becomes mud, it is too wet. The dirt mixture should be one to two inches deep. It is very important to add peat/soil whenever the medium becomes too wet, but also if it begins to have anything but a fresh odor.

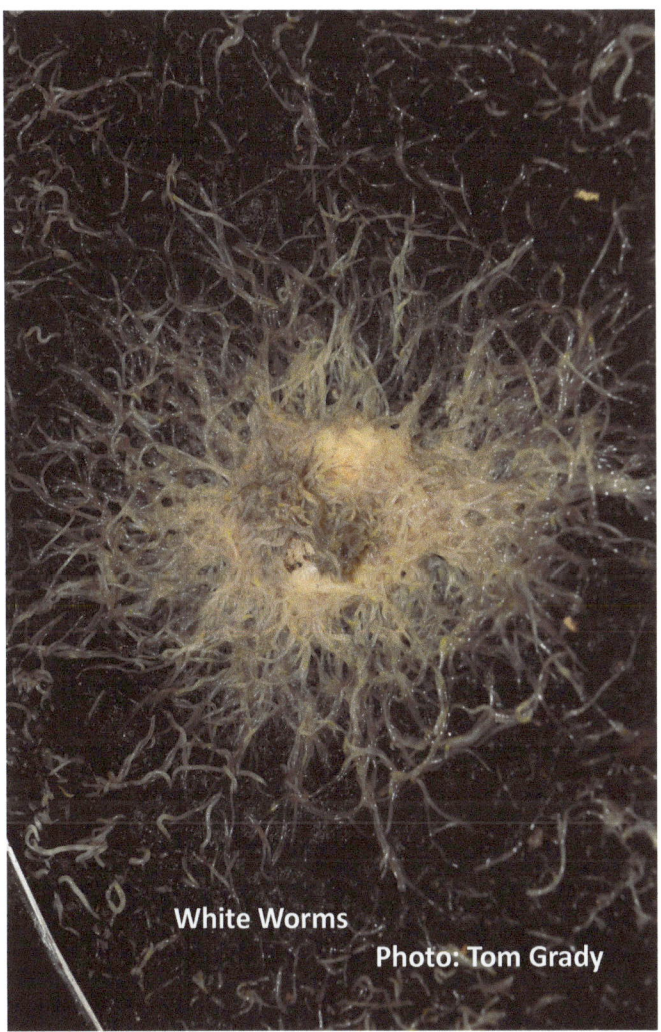

White Worms
Photo: Tom Grady

It should be noted that most forms of white worms prefer a temperature range of 55F to 65F for optimum reproduction. At higher or lower temperatures the production trails off.

There are a number of containers that can be used to house the culture ranging from wooden boxes to styrofoam. The benefit of wood over the styrofoam is excess water in the soil tends to be absorbed into the wood and then evaporates. The best location for the containers is a place that is dark most of the time.

The soil mixture in styros needs occasional tending with the addition of more peat moss. Of course the benefit of adding more media is the hobbyist can split the culture in half and start a new one easily. When starting a new culture from scratch, it normally takes one to two months for it to become fully established and producing enough worms to feed the fish. This is relative, of course, to how many fish are being maintained.

White worms need food to grow and reproduce. Any number of 'organic' foods, including Cheerios, mashed potatoes and dry cat food, but simple white bread soaked in yeast water is the least expensive and works quite well. The benefit of cat food is the additional protein causes the worms to reproduce faster. Mashed potato flakes can simply be placed on top of the soil for a quick feeding. They will absorb some moisture as a side benefit and then expand.
There is nothing wrong with mixing any or all of these foods together to provide the best overall nourishment.

White bread: The addition of fry active yeast to water helps minimize fungal growth on the bread while the worms consume it. The yeast should be soaked in warm water until it becomes milky. Supplements, vitamins and vegetative matter can be added to the water. Once the mixture is ready, the bread should be soaked and introduced into the white worm culture. (Alternatively, the use of a sprayer

KN PRESS - T.R. GRADY

once the bread is placed will provide the same results.)

Place the bread on top of the soil and the worms will find it. Some hobbyists will place a piece of glass or clear plastic over the bread. This probably is used for two reasons: to keep the bread moist and/or to allow the worms to gather beneath and around the bread. Often balls of the worms will gather beneath the food and are easy to remove. For the most part, worms will cover the top of the food and many will attach to the glass for easier harvesting. Other aquarists place the bread on top of the glass and the worms will climb onto it to reach the food making them very easy to harvest. This has the advantage of less soil debris and waste detritus that needs to be cleaned.

To harvest the worms all depends on how squeamish the hobbyist might be. Clumps of worms can be removed simply by fingers and placed into a container of water. Some people will prefer not to touch the worms and use a pair of tweezers. It really does not matter how the worms are removed from the culture, but cleaning them before feeding the fish is important. Whether you are removing waste byproducts or trying to remove excess peat/soil from the culture the worms need to be well rinsed. The use of a four-inch deep plastic container (such as one used to package deli foods - potato or macaroni salad - at your local supermarket) allows for the running of tap water into worms and then vigorously stirring the water before pouring it carefully off to remove the unwanted wastes. This should be done several times until the water runs clear. For those interested, careful examination of the water will reveal some very tiny worms. These can be poured into another container and will settle to the bottom. They can then be siphoned off and used for the smallest of fry.

Once white worms are clean, the easiest method for feeding them to the fish is with the ubiquitous turkey baster. Again, depending on how many tanks of fish need to be fed, the size of the water container is optional. Anything from a wide-mouthed plastic peanut butter jar to a salad container may be all that is needed. The worms are placed in the container, the container filled with fresh aquarium water (based on need) and then stirred with the baster and drawn up. Spray the worms into the tank and the fish will be quite happy.

Grindal Worms

Grindal worms, *E. buchholzi,* are a smaller species of segmented worm and make a superb and relatively easy to culture food for smaller species of killifish as well as juveniles. While Grindals can be cultured in an identical fashion as white worms, there are easier techniques. At least three methods are in common use today and hobbyists have created many variations of those. All of the procedures work equally well in maintaining a constant source of food. The Foam Sponge Method and the

Note: Siphon off the minuscule white or Grindalworms and add them to the baby brine shrimp container which is used to feed the fry. This gives the fry an option as to the food they prefer at that moment. This can also be done with vinegar eels and microworms. For adult fish, frozen brine shrimp and/or live Daphnia can be added to the mixture and all fed at one time.

Green Scrubbing Pad Method are described below. It all depends on the hobbyist's preference. The third method, just to clarify is the same as White Worms.

Although Grindals are smaller than white worms, they tolerate a higher temperature range and can be cultured at room temperature.

The Foam Sponge Method:
Parts:

1. **Plastic box** - this can be a variety of sizes but needs to be at least 1-inch deep. There are a number of options including shoe boxes, sweater boxes and plastic containers for a variety of needs. They all require a tight fitting cover. They may be clear or opaque.

2. **Glass** cut to fit tight inside the box.

3. **Sponge** - This should be a small pore foam pad, often sold in department stores or fabric shops as filler for pillows and cushions. It should be about one half- to one-inch thick depending on the size container being used. It should not reach the cover of the box

The sponge can be cut to fit tight to the sides of the box for small containers, but this is not necessary for larger ones. The glass should fit over the majority of the sponge. Water is then added to the system until it is absorbed about half way up the sponge.

To Inoculate a New Culture: The starter culture of Grindal Worms is placed in the center of the sponge and a few pellets of food such as kitten chow are added where the culture is inoculated. This is covered by the piece of glass and placed over the sponge to hold in the humidity. The kitten chow will absorb some of the moisture in the foam, soften and break it down. This is what the worms feed upon.

It will take a bit of time to have a well producing culture, perhaps as much as a month. Once there are enough worms to use as part of the feeding process, simply remove the glass from on top of the sponge and 'scrape' off the worms into a water filled feeding container.

Some residual food may remain on the worms, but once in the fresh water and stirred it is minimized and will not affect the water quality. The worms can be drawn into a turkey baster and added to another fresh water holder for additional pollution reduction and the fed to the tanks.

To maintain the culture, add new kitten chow on a regular basis. To determine how often this needs to be done, simply wait for the older food to be fully consumed. Be aware if you use too much, it will begin to form mold and must be removed.

It is absolutely necessary to keep the culture clean. The water in the container should be drained at least once a week and replaced. Tip the box and squeeze the foam several times to remove the brownish water that contains the waste products of the worms. Pour this water down the sink. Replace with fresh water and then repeat the process as many times as necessary until the water runs clear.

The Green Scrubbie Method:

This is not much different than the Sponge Method, but the primary culture media is green scrubber pads placed inside a plastic container. These pads are readily available in grocery and department stores. A small amount of water needs to be maintained beneath the pads, covering the lowest one. The water is drained on a regular basis to remove the wastes. The pads can also be carefully

KN PRESS - T.R. GRADY

rinsed under tap water, taking care not to wash out the worms, for the same reason.

Generally four to five pads are stacked to allow the culture to become established. A small amount of a starter culture is placed on top of the pads and food placed on top. The use of kitten chow is the preferred nourishment Some types of kitten chow will grow mold much faster than others. Purina One Kitten Chow is one of the better choices.

As the culture grows the number of pellets can be increased. The key to this is the 24-hour rule. If the food disappears in 24-hours, then the correct amount of food is being fed.

Harvesting the worms is probably best done with a rubber-style spatula. Put the removed worms in a container of water and stir them up. Allow the worms to drop to the bottom, pour off the cloudy water and repeat. This will rinse the worms of any byproducts not welcome in the aquarium.

If desired, a piece of glass or plexi-glass can be used to cover the pads and the worms will attach to it and are easier to remove for feedings.

The question of using a sealed cover only depends on the hobbyist. The worms generally will not climb out of the culture, although it is possible if the media becomes contaminated they may well climb the sides in an effort to escape. If a cover is used, it is important to make certain there are numerous air holes in it.

The one concern with any live culture is having it become infested with mites and/or fruit flies. The addition of a small piece of "No Pest Strips" will eliminate this issue and not harm the worms. These yellow strips can be purchased in most large box stores. This is a recommended addition to any live food containers for white worms and Grindals.

Vinegar Eels

Vinegar eels are one of the simplest live foods to culture for baby and young killies. The only negative is you may notice the odor of vin-

egar whenever you expose the culture to the air. Vinegar eels also have the advantage over microworms in that they will swim through the water column to the surface while microworms remain on the bottom of the tank.

To produce these tiny worm-like eels, a starter culture is necessary, but can be obtained from just about any breeder or on-line from a variety of sources (hobbyists, scientific supply houses and Aquabid).

The culture media that works well is a simple mix of apple cider vinegar and clean tank water. Use this at a ratio of 3:1 - three parts water to one part vinegar. A quart sized container works quite well and lasts for months. Add the starter eel culture to it and then wait until it is established. Essentially this means it is easy to see clumps or balls of eels gathering at the surface, generally along the edges and in the corners of the container (depending on the shape.) It is simple to harvest the eels by using an eyedropper and then adding them to a larger container of clean water. If careful, the amount of actual vinegar added to the feeding vessel will be so small as to be negligible. For example, your initial mixture is 3:1 parts water. Using a 'cup' sized container of water for feeding means the eyedropper amount of vinegar is now something around 1000:1 parts water.

The Eels can also be harvested by leaving a green scrubbie in the container. Eels will get into the scrubbie and after it drains, place it in water and squeezed to release the worms. There will still be a bmall amount of vinegar in the mix.

Vinegar eels can be seen by eye, but using a small magnifying glass eases the stress to view them. Essentially the hobbyist will see tiny white 'worms' wiggling in the water column. For feeding baby killies, simply add a small amount of the eel bearing water to the fry container. The eels will swim to the surface where they are easily found by the fish. Since we all should be doing regular, even daily water changes for young fish, any residual vinegar becomes far more diluted and does not pose any problems.

Some hobbyists will mix vinegar eels with newly hatched brine shrimp nauplii to create an easy feeding of the two foods at the same time.

Microworms, Walter Worms & Banana Worms

Microworms are a tiny nematode excellent for feeding newly hatched killies or any other small fry. They are easy to culture and only require a three to four week restart of a new culture to maintain a continuous source of nutrition. Some hobbyists maintain several of these cultures at all times simply to keep up a rotation and provide enough worms for feedings to large fish populations.

A variety of containers, all small and clear are available for use. Once again the small clear deli container is perfect. In this case the top in also needed. Several holes need to be made in the cover to allow air into the culture. Cultures can be kept in a refrigerator to slow the speed at which they begin to sour.

There are a number of cereal mixes that can be used, but oatmeal (some people prefer the baby food version) works fine and is a cheap and easily available product. Simply add water to create a slurry or paste and pour this into the container to the depth of about 3/4 of an inch. Sprinkle the trusty dry active yeast over the top of the mixture. The yeast are what the microworms feed on. Now add the starter culture of microworms. The shimmering on top of the culture media indicates a healthy growth of the microworms. Within a few days microworms will begin to climb the sides of the container and may be fed to the fish.

Using a rubber spatula, scrape them from the sides and introduce the worms into a water-filled feeding container. This can be the same container used for baby brine shrimp to offer the fry options when feeding.

In three to four weeks, the culture will begin to have an odor. This is the indication it is time to start new culture. Simply create a new container of the chosen media exactly as described above and take a small amount, about half a teaspoon, of the old culture and spread it across the media. Within a few days a fully functional culture will be established.

Variations of Microworms

Walter Worms: These tiny worms are another species of nematode that is smaller by half than microworms. Culturing them is identical to the microworms technique. They tolerate higher temperatures than their larger cousin. They also live for about 24 hours inside the fish tanks, making them less likely to affect the water quality.

Walter Worm cultures generally last about 10 days before they begin to go bad. A vinegary odor will begin to emanate from the culture and this indicates it is time to start a new culture. Like microworm cultures use a small amount of the dying culture to inoculate it.

Banana Worms: Even smaller than Walter Worms, Banana Worms have become more popular recently for newly hatched fry. They can be cultured identically to the two previous worm cultures, but some people add mashed banana to the media mixture (hence the name Banana Worms). This live food can tolerate warm temperatures up to 85F. Normally the life span of the culture is about 10 days, so new cultures need to be made on a regular basis. Alternatively, additional culture media can be added to the container to extend the life of it a short time.

Black & Tubifex Worms

These nutritious worms are both an excellent source of nourishment for killifish as well as a potential source of disease and other problems. There is probably no better live food for conditioning females for breeding.

It is not likely anyone is culturing them as a food, but they can be purchased on-line, generally in increments of pounds, and shipped to the hobbyist. What is important is how they are cared for once received.

First, they must be kept cool. Most of these type of worms are found in cool running water and are often common on the fringes of larger fish hatcheries. They produce considerable waste and cleanliness is essential.

A refrigerator is required, although if a hob-

Most experienced killie-keepers prefer Black worms over Tubifex. Many cite that Tubifex are much more likely to bring disease into the tanks.

byist wants to make the investment, a continuous flow-through system can be built and tap water used to continually rinse the worms.

In the refrigerator, an opaque sweater box works well as a storage container. Enough water is added to the container to just cover the ball of worms and the cover is loosely placed back on the box. Because of the wastes and also dead worms, the water needs to be replaced on a regular basis, at a minimum daily, and the worms rinsed well. To keep them clean, it may take two or three complete water changes a day. Make certain the water flows clear before ending the rinse procedure.

The container is now placed in the refrigerator. The cool temperature slows the worms and helps minimize the wastes.

To feed the worms to killies is no different than the techniques used for other worms. Place the worms in a feeding container filled with clean water and then use a baster to spray them into the tanks. Try never to feed more worms than the fish will consume in a few minutes. Having them die on the bottom can lead to fouled water.

Earth Worms

Red Wrigglers - earthworms, -are a great food for larger killifish including the Blue Gularis, *Fp. deltaense* and a variety of other predatory species. They are not difficult to culture, but will take up a little time to set up and some space in the fish room. The hobbyist will need to obtain a lightweight bin, generally a thirty-gallon plastic container that is water tight, and allows limited light to penetrate.

The top of the bin will need several small holes drilled in it to allow air flow as well in the upper sides.

The Bedding: Worm bedding is available almost anywhere fishing supplies are sold and can often be found in large box stores. A variety of alternative options including peat moss, coconut coir and shredded paper, newspaper or cardboard will all work as a place for the worms to live and reproduce. Avoid glossy type papers such as magazines.

Food: Nearly any organic food scraps can be used for the worms, but lettuce, apples, carrots, cereals and even mushrooms are all readily taken. Even coffee grounds and old teabags are useful. Avoid meats, dairy products and sauces. As the culture reaches its maximum size, the worms can eat up to 3 pounds of food a week, so be prepared to provide a good portion on a regular basis. The better chopped up the food, the easier it is for the worms to eat.

It is suggested to layer the bedding with the food and then add a second layer of the bedding with more food on top and then a final layer of the bedding. Garden soil or peat moss should be added on top of the bedding. The worms need the microorganisms in the dirt to help process the food. Well dampen the upper layer with about one to one-and-a half liters of water. Six or so crushed egg shells should be added and then it is all mixed together.

At this point the starter culture of worms can be added.

Depending on the needs of the hobbyist, the worms can be harvested easily by hand if only a few are used on a regular basis. If larger numbers of worms are needed, a neat little trick would be to place a handful of the media on a large white piece of plastic and left under a bright light. Since the worms are light sensitive, they clump together and can be removed in a large group to be washed.

One of the benefits of this process is the castings of the worms is an excellent fertilizer for gardens. For a bin to be converted to castings takes about 5-6 months. For serious hobbyists, starting a new container of red wrigglers every couple of months will pretty much guarantee a continual food source.

Feeding earth worms to your fish: Depending on the size of the fish the worms are intended to feed, very small ones can easily be taken whole, but larger worms should be cut up to make it less likely to choke the fish. The worms themselves need to be cleaned. There will be detritus/wastes from the castings on the skin of the creatures and most likely waste products in the intestinal tract. There is no simple mechanism for the worms to be cleaned and the best seems to place them in a container of water for some time. As the dirt dissolves off

the sides, the worms become pink and appear clean. Internally, hopefully, they will expunge most of the wastes into the water. Rinse them a couple times to remove any residual wastes and then add them by hand to the tank with the awaiting diners.

Fruit Flies (Wingless)

An excellent food for tropical fish, wingless fruit flies (actually have vestigial wings) are also the bane of the hobby when they escape the self-contained cultures. Once fruit flies discover freedom, they find any number of places to reproduce and you may well need to fumigate your home to get rid of them. Without a doubt this should seriously be considered before adding them to regular feedings. On the positive side, many species of killifish absolutely love them as a source of nutrition.

While there are a number of species available to culture, probably *Drosophila melanogaster* is the best choice for killie-keepers. This species grows to about 1/16th of an inch and is small enough for most killies to eat. *Drosophila hydei* is a larger fruit fly, obtaining 1/2-inch in size and may well be an option for hobbyists maintaining larger killies such as *Fp. sjoestedti* or some of the larger carnivorous annuals.

Creating cultures for any fruit fly is identical and not at all complicated. The container should be fairly large. Mason jars, mayonnaise jars, quart-sized glass containers, peanut butter plastic containers and quart-sized deli containers all work. A cover to prevent any of the flies from escaping is absolutely necessary and that cover needs to be ventilated. A variety of ways to accomplish this are available, but in general using a coffee filter over the opening and sealing it in place with a rubber band or the actual rim of the original metal/plastic cover works best and is simple (a hole can be cut in the top). Some hobbyists have used cotton stuffed into holes in the cover. Whatever method is chosen, unless you want your spouse to suggest you find a new home, be aware of potential escapees.

The use of sticky 'fly tape' in the area near the cultures helps immensely. "No Pest Strips" can also be placed nearby the containers and not harm the cultures. These strips are exceptionally good inside almost any worm culture to eliminate both fruit flies and other pests from becoming established.

The culture container needs to be cleaned throughly to minimize the possibility of contamination because the food you use will create an environment where bacteria could thrive.

A good homemade recipe for feeding the fruit flies would be a mixture of four parts mashed potato flakes, 2 parts powdered milk and one part sugar. Add an equal amount of water to create a thick paste of oatmeal consistency. To this sprinkle some dried active yeast over the top and allow it to stand for a few minutes. When ready, spread this mixture over the bottom of the container.

The flies need something to lay eggs on and grow. There are any number of options ranging from popsickle sticks, aluminum screen to plastic meshes that can be modified to fit inside the container. It is suggested these be angled from one edge of the container to the opposite side and the lowest end stuck in the media on the bottom.

Room temperature works fine for these cultures and they should last four to six weeks.

Depending on the needs of the fish room, new cultures can be regularly started daily, weekly or monthly by simply inoculating them from an active one.

Feeding them to the fish tanks is a little more difficult, only because this can be a chance for them to escape. Initially the hobbyist should shake all the flies to the bottom of the culture before opening the top. Quickly tap out a few flies into the water and then close the top again quickly. As long as your tanks are covered, it is not as likely many will escape the hungry fish.

The alternative to this is to purchase fruit fly kits on the Internet which come with food, containers and starter cultures. Some sources for this are listed in the resource appendix.

Infusoria: The Micro-Foods

Paramecia, Euglena and *Rotifers*, all forms of infusoria, are nearly microscopic animals which are among the best starter foods for new hatched killifish. All are mobile and the movement attracts the fry. While the term infusoria is actually obsolete, it has become the staple of the tropical fish hobby to denote the minute creatures. There are literally thousands of forms of infusoria and many can be easily cultured for use in the home aquaria environment.

Realistically, infusoria are present in every established aquarium, normally hiding in floating plants, but actually everywhere that they can crawl or swim. The numbers are enough to sustain a number of fry if hatched out in a community environment until they grow large enough to feed on other foods. However, not enough of them will be available if fry are being raised in numbers in a controlled environment and then must be supplemented by a maintained culture.

In general, a hobbyist could initiate an infusoria culture without needing a starter culture, but to use specific forms of the microorganisms, a starter is required. To start a culture all one needs is a jar of aged (tank) water and some form of decomposing vegetative matter such as papaya skin, lettuce, or banana skin. Hay has also been used. Place the jar in a window or a bright spot in the fish room and wait a few days. When the water becomes cloudy, that means bacteria have formed and is what the infusoria generally feed upon. The exception to this is Euglena which has a mechanism which allows it to produce chlorophyll. Since tank water is used, there should already be infusoria in the water column and all that is needed is to allow it to reproduce.

Once a culture has been established, it does not hurt to add a few snails to clean up the remains of dying microorganisms as well as produce additional bacteria with their wastes. The snails will also feed on whatever decomposing vegetative matter is added to the culture.

The following three forms of infusoria are all superb starter foods for killie fry as well as any other species.

Paramecia: *Paramecia*, a protozoan, are a good first food for all types of small fish fry. While they are nearly invisible to the human eye, the fish see them easily. If a hobbyist looks very closely at a culture, tiny specs that seem to float like a cloud are most likely the paramecia.

To maintain a culture of paramecia is not at all difficult, but it does take a little care. Start with a sterilized gallon jar. It can be sterilized by pouring boiling water into it. Once cleaned, fill the far about two-thirds with fresh dechlorinated water. You want to leave as much surface space as possible. Tank water is not a good choice as some microscopic creatures always exist in older water, whether transported there by plants or some other means.

The jar should be kept in a dimly lit location. Some sort of vegetative matter needs to be added to the jar, dried corn husks are a good

choice, but dried lettuce leaves, even kitten chow and many other dried products work just as well. This provides a reason for bacteria to become established. The *paramecia* feed off the bacteria.

Now add a starter culture of the paramecia. If everything went correctly, then in a little less than a week, the culture should be well established and a cloud of *paramecia* becomes obvious. These cultures do not last for a long time and peak at a bout a week. It is wise to start a new one every few days.

To feed the fry, simply remove some of the cloudy water from the jar with a baster and add a few drops to each fry tank.

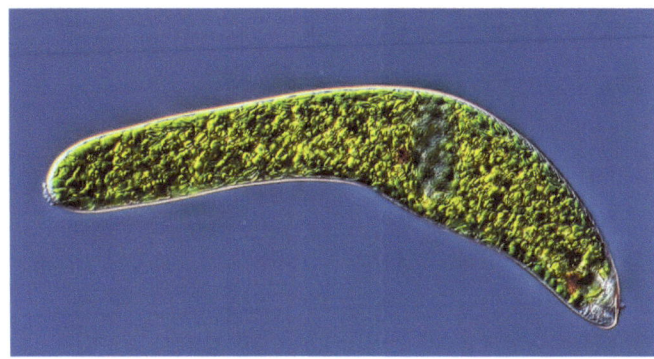

Rotifers: *Rotifers* are another excellent source of food for tiny killifish fry.

However the negative aspect of *rotifers* is they are best cultured in salt or brackish water. The salinity can be half of that used for a marine aquarium, so that helps with both cost as well as feeding. Technically they can be grown at lower salinities. Room temperature works fine. A specific gravity between 1.00075 and 1.-26 (about 15-20 ppt) works best.

Perhaps the easiest method of culturing these microorganisms is to use 3- to 5-gallon plastic buckets. A flow of air is necessary to circulate the water and a fine mist air stone is probably the best option. Fill the bucket with your salt water mixture.

You will need a microalgae feed such as RGComplete, which is available at APBreed or culture your own microalgae. Microalgae is necessary for the rotifers to eat.

Once the buckets are ready, simply add a starter culture and wait for at least three days to allow it to become established.

Once established and you are ready to feed, turn off the air, stir the culture vigorously and wait for five minutes to allow the green detritus to settle to the bottom. The rotifers will swim to the water's surface. Now you can harvest them by using a baster and removing them from just below the surface. The baster of rotifers can be added to a fresh water feeding container and then sprayed into each fry tank. Alternatively, the rotifers can be added to a mix of baby brine shrimp, paramecium and/or vinegar eels.

Euglena: *Euglena*, better known as green water, is a very different animal. To qualify the term 'green water' it should be understood it is also used for algae. It is necessary to understand the green water we want is actually *Euglena*.

This family of single-celled flagellates is interesting because they are capable of photosynthesizing chloroplasts within their bodies. This is how they are able to feed themselves, much like plants. It is not the only way they feed however. They are also capable of surrounding a particle of food, such as bacteria, and consume it through (simplified description) absorption.

Euglena reproduce through binary fission, the division of cells. Essentially the nucleus of the cell performs mitosis and then the Euglena forms the internal mechanisms (flagellates for movement, gullet for feeding and stigma) for survival before separating into two individuals.

A simple method of culturing Euglena is to use a large glass jar (sterilized pickle jar) and add clean aged water. Some sort of microalgae nutrient (See appendix for sources) is valuable and a starter culture which can be obtained from existing cultures, from other hobbyists or scientific supply houses. A strong light source is necessary since *Euglena* photosynthesize most of their own food. However, direct sunlight will cause the water temperature to rise above a level in which the Euglena cannot survive.

Euglena will settle to the bottom of the jar and form a 'scum carpet'- actually a colony. It does not hurt to stir up the culture (or shake it) to keep the organisms suspended in the water. Once the water turns green the culture is fully active and may be used to feed fry. It is important to regularly start a new culture in order to protect against a complete loss.

Once you are producing large numbers of *Euglena*, as well as feeding newly hatched killies, it is an excellent source of nutrition for *Daphnia*.

Health Care: Disease & Infestations

Like any group of tropical fish, disease can be introduced into the fish room, somehow spreads and wreaks havoc. This most often happens when new fish or plants are introduced into the system, but some of the issues can be caused by poor water quality. Most diseases that infect killies are no different than the ones infecting all tropical fish, but there are a few that are of particular interest for killifish hobbyists. Some deadly conditions seem to present themselves in killifish primarily and are rarely seen in other species. These are addressed in their own section.

Prevention, Prevention, Prevention

Prevention is always the best option. Whenever new fish are introduced into a system, there is a good possibility they have already been infected by something, even if no symptoms are present. These fish have faced considerable stress in transport, whether it is from another aquarist or purchased in a tropical fish store. Imagine what wild fish are subjected to and how many unknown diseases might be present.

Just the process of bagging a fish can cause own section following the identification of the more common conditions.

White Spot Disease or Ich:
Ich or *Ichthyophthirius multifiliis* (White Spot Disease) is actually a protozoan infection. One of the most common conditions in the tropical fish hobby, it is also found in killies.

Considered an ectoparasite, it forms small white nodules on the sides and fins of the infected fish. It can be very damaging to the skin and gills of the fish. Uncontrolled it will kill.

Ich goes through three life stages once established in a tank. The first is a feeding stage which presents itself as the white spots cov-

enough stress by damage to the scales or natural slime coat and that can allow a parasite to attach itself. If the fish have come from a fish store, or even worse a large box store with a fish department, those tanks most likely are already contaminated by a variety of diseases, parasites and/or other problems. Any or all can now be added to the hobbyists' system. Not all diseases manifest themselves until an appropriate host - weakened or damaged fish - presents itself.

The first line of defense in preventing an infection is to isolate newly obtained fish in 'hospital' tanks. This is particularly essential for species intended to join larger established communities, whether it is a multi-species system or simply new blood added to an established breeding colony. Once the new fish have spent a few days isolated, any likely symptoms should appear and can be treated.

Dead Fish:
Water quality is probably the single most important factor in maintaining healthy fish. While this was discussed earlier, it is very important to state dead fish must be removed as soon as they are detected. As the bodies begin to decay, ammonia enters the process and then affects the overall system. It does not take long to contaminate the tank. The fish sicken and eventually could die.

Common Aquarium Diseases

All tropical fish are susceptible to a myriad of diseases and conditions and killies are no different. Most of the common illnesses and parasites can be as much a problem in the killie hobbyist's tanks as they can be in home community tanks. However there are a couple of problems which are of particular interest to killifish keepers. These are detailed in their

ering the body and fins. The second is when it forms a trophozoite and falls off the fish to the bottom or any ornament including plants. This becomes an encapsulated dividing stage called a tromont where it can adhere to anything in the tank, including nets introduced into the water column (which can lead to a transfer of Ich to other tanks.) The tromont can divide up to 10 times which then enter a free swimming stage. They now can attack a new victim and the cycle repeats itself approximately every seven days.

Treatment: There are a number of commercial treatments which work to a degree as well as the use of salt under certain conditions. Most of the chemical treatments are focused on killing the parasite in its free-swimming stage and recommend the rise in water temperature to around 80ºF (27ºC). This is to assist the life cycle of the Ich to progress rapidly and does not actually affect the protozoan although some claim a temperature of over 86ºF (30ºC) will kill the Ich in the free-swimming stage. Many of these treatments include a dye (malachite green) of some sort as well as a salt such as zinc and copper.

Once the Ich is eliminated from the tank, it is wise to do large water changes over a few days to remove any chemicals and dyes that remain.

Fin Rot: Fin Rot is a condition caused by a bacterial infection, generally *pseudomonas*, but can be caused by other gram-negative bacteria. It is initially identified by a white or milky clouding along the edges of the fins. At this point the progress can be halted relatively easily. If left untreated, the fins will slowly become frayed and become inflamed. Tissue will drop off and give the fins a ragged appearance. Once it reaches the body, ulcerations form and it is unlikely the fish can be saved. Before that secondary infections may take hold, *Columnaris* being one of the more common. Once *Colum-*

naris takes hold, the fish is doomed.

Fin Rot is initially a reaction to stress which can be anything from poor water quality to shock from temperature and rapid water condition changes, including changes in pH or water hardness, that cause the slime coat that protects the fish to dissipate. Overfeeding is also a factor.

Treatment: Treatment consists of the use of one of several commercial tropical fish antibiotics. It is absolutely necessary to make water changes to eliminate the original reason the bacteria appeared. The use of aquarium salt benefits many species. Prevention is always the best option and good aquarium maintenance will reduce the possibility of infection.

Fungal Infections: There are two types of fungus killifish hobbyists deal with. The one in this discussion is body and fin fungus. Egg fungus will be discussed in its own section in the Killifish Egg section under Breeding Killifish.

Fish Fungus is an opportunistic condition that takes advantage of injuries or damage from other conditions (Open wounds caused by Ich, hole-in-the-head disease, or ulcers). Fin-nipping can be another issue that leads to infection. Species of *Achyla* or *Saprologia* are present in most aquariums as part of the breakdown process of uneaten foods and detritus. Given the opportunity fungus will take hold in the sores and even the living flesh of the fish. This infection will appear like cotton patches.

Treatment: While fungus is not considered communicable, it is still wise to move the affected fish to a quarantine area for treatment. Fungus expands rapidly on the infected fish and treatment must be started as soon as possible. Organic dyes like malachite green and acriflavin combined with formalin are most commonly used as a treatment. Commercially available variations of this treatment can be purchased from a variety of sources.

Cleanliness of the ecosystem is considered the best preventative medicine in the case of fungus. Water changes and good filtration are important. The removal of organic matter from the aquarium is essential to good health care.

Another infection with a similar appearance is **Columnaris,** a bacterial, not fungal, infection commonly found on the mouth, but also can attack the fins. It is caused by the gram-negative bacterium *Flavobacterium columnaris*. It is often misidentified as a fungus, but is far more deadly and tends to begin in the mouth of most fish. This condition is highly contagious and must be treated as soon as identified. As the infection progresses, it covers more and more of the affected area and will engulf an entire fish if given the chance. Beneath the cottony covering, ulcerations form and become necrotic. This will happen in the gills also. Fatality generally occurs with 48 hours.

Treatment: Realistically, if the infection becomes established it is unlikely the fish will survive. There are antibiotic treatments on the market including tetracycline variations, kanamycin and furan-2 and are used in conjunction with the ever-popular malachite green or methylene blue.

Internal Parasites & Worms: A number of worms have been known to infect killifish over the years. Probably the most common are the numerous *Cammallanus* species, but anchor worms also occasionally appear.
Camallanus worms generally appear as red threads protruding from the anus of aquarium fish. This coloration is because the mature worms feed on blood. Other symptoms include avoidance of feeding, wasting and abdominal bloating due to the irritation of worms feeding which can lead to secondary infections in the gastrointestinal tract.

Camallanus worms go through three stages,

only one which leads to the infection of fish. An infected fish will contain pregnant female worms which then reproduce and create large numbers of larvae. The larvae pass from the fish via waste materials, exist in a free-living stage in the substrate where the first in a series of molts takes place. During this molting stage the larvae infect an intermediate host such as *Daphnia, Grammarus* or *Cyclops* where they continue to molt twice more and then enter an inactive stage. These crustaceans are then eaten by the fish, another molt takes place inside the new host and the third stage larvae begin to feed on the fish. The process takes less than a month at room temperatures.

Treatment: The primary treatment for *Camallanus* worms is the use of Antihelminthic medications. Several of these are available in the aquarium trade including Fenbendazole™, Flubendazole™ and Levamisole™. These drugs attack the infestation internally. It is known that some of these drugs only paralyze the worms and they frop out of the fish to the bottom. Removing the worms becomes necessary.

Anchor Worms: are a copepod crustacean which can infect tropical fish and appear as reddish or green-white thread protruding almost anywhere on the fish. They are easily seen by the naked eye. Affected fish are often seen rubbing themselves against the gravel or substrate. There will be a localized red inflammation at the spot of infection. The fish will become lethargic and develop labored respirations.

Sadly, the best treatment option is to remove the Anchor Worm by hand (tweezers) and then treat the wound site to prevent other infections. Potassium Permananate is recommended as a post treatment additive as a preventative. Clout™ has been stated as effective by some and commercial companies use a product called Dimillin.

Fish Tuberculosis: Tuberculosis in tropical fish is cause by the bacteria *Mycobacterium marinum* and is closely related to human tuberculosis. It is one of the most difficult conditions to treat and it can be spread to humans. If fish TB is suspected, the appropriate precautions including the use of long gloves must be taken. Avoidance of the water is important.

Symptoms: Weight loss and emaciation, scale loss, sores and lesions and skeletal deformities.

Treatment: Euthanize the fish. Sterilize the tank.

If a hobbyist suspects Fish TB has been contracted, an immediate trip to the doctors office is important. Be certain to mention the possibility of a fish related disease. Symptoms initially include a rash that progresses to small purplish lesions. Untreated it can enter the joints and bones and do considerable damage.

Killifish Specific Conditions

While all of the following diseases and conditions can affect most species of tropical fish, they are often found in killifish and deserve a specific identification for a variety of reasons as described under each.

Glugea or 'The Notho Disease': *Glugea*
is a microsporeum infection of the intestinal tract of some killifish and has been extremely destructive in species of Nothobranchius in the past. It has been identified specifically as *Glugea anomala,* and is known to infect other species of fish including sticklebacks.

For the hobbyist, it has been observed primarily in certain annual killies including many species of *Nothobranchius* as well as *Austrolebias* and *Fundulopanchax*. It is likely capable of infecting most species of tropical fish if they are exposed.

As a condition it presents itself initially as if the fish are very well fed and forming rounded bellies. The fish initially act normally, but continue to become fatter. This is actually the microsporidium reproducing and expanding exponentially in the intestinal tract. At some point, either a protrusion will extend from the anus of the fish, or the side/belly of the fish will erupt in a white sore or numerous sores.

These fish should be destroyed immediately. There is no cure for them at this point and the goal now is to prevent the spread of the disease process. It is absolutely necessary to completely sterilize the fish tank. This is best done by filling the tank with chlorine bleach and allowing it to stand for several days. After the tank is re-emptied, it should be allowed to dry out and remain unused for a period of time up to one month before adding water and fish to it.

How *Glugea* Spreads: The most important thing to understand is *Glugea anomala* reproduces via spores released into the water column, settle to the bottom and are ingested as the fish pick through the substrate seeking food. These spores are released along with waste byproducts as well as through the open sores on the body and will settle into the substrate. The cysts are nearly impossible to destroy. For keepers of annual killifish, the spores enter the breeding materials (peat moss, coconut fiber, etc.) and then enter a dormant stage when the media is processed and dried for storage.

When the peat moss is placed in water for hatching, the spores release the infection into the water column and the next generation of fry become quickly infected. It takes six weeks to two months for the symptoms to initially appear and the hobbyist may not be able to recognize it until well after the fish are already spawning.

Unfortunately one of the most serious problems of *Glugea anomola* comes when killie-keepers sell and trade the bags of peat moss around the world. This can spread to infections becoming established elsewhere before anyone recognizes the problem.

Early Treatment: Hobbyists have found that Flubendazole™ may be used to treat this disease somewhat successfully in the early stages as well as a preventative. One technique that is practiced is to add the drug to the surface of the water when hatching bags of peat moss. This will hopefully inhibit the infection of the baby killies. Packaging suggests using it with foods and to soak it for 30 minutes. It is not very soluble in water, but if used while feeding, the fish may well ingest some of the particles. Flubendazole™ is an anthelmintic that used by veterinarians against internal parasites for larger animals. It is available commercially for the aquarium trade.

In addition, at the earliest stages Flubendazole™ can be added to infected tanks and hopefully the fish will be able to ingest enough of the medication to enter its intestinal tract and kill off the internal infestation. Realistically, the best option is preventative.

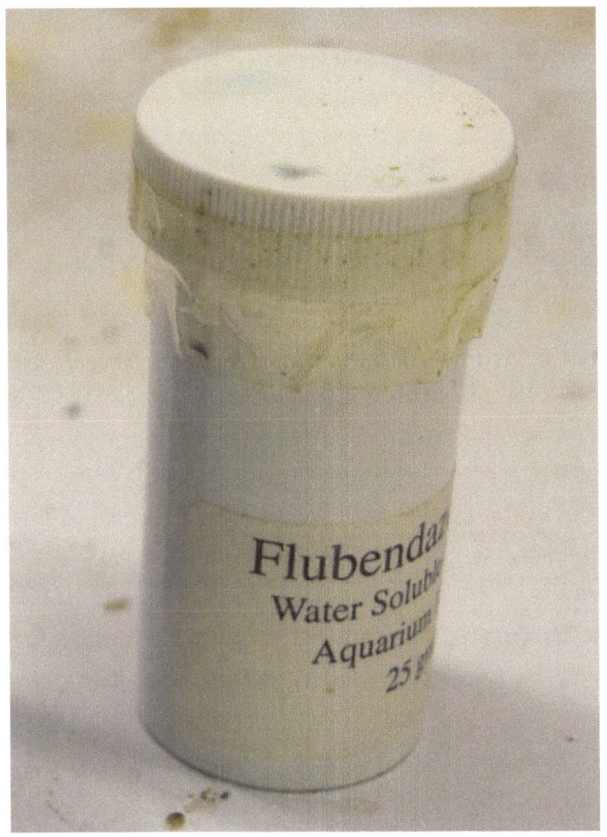

Dropsy and Bloat: Occasionally an aquarist will see a fish that becomes bloated and the scales seem to stand out on end in a 'pine-cone' effect.. This is Dropsy, but may be known as Bloat of even Malawi Bloat. It is actually a symptom of a bacterial infection most likely caused by poor water conditions, but primarily affecting the kidneys of the fish. This causes edema, or the retention of fluids in the body tissues.
High nitrates seem to play a part.

It is very difficult to cure and the preferred course of action is to destroy the infected fish. While it is not considered contagious, the conditions that led to the fish becoming infected need to be addressed. Antibiotics may have some affect on the disease, but it is rare a fish in this condition does not die. Since it is internal, the medication needs to be ingested which leads to soaking the food in a liquid containing the antibiotic. On rare occasions the fish will recover on its own.

Prevention is simply maintaining good water quality in the aquariums, siphoning the detritus and doing water changes.

Genetic Defects: In-breeding is a problem with any isolated population, whether in the wild or the home environment, of fish, killies included. The genetic diversity becomes locked and disorders appear. Some species lose vigor and ultimately fail to reproduce.

Many species of killifish have been introduced into the hobby and a few generations later, are gone because of a refusal or inability to reproduce. Other issues include physical deformities such as curvature of the spine, deformed fins, maximum growth becomes smaller or the original colors diminish.

Unfortunately this is a product of the hobby in general. Many times only a few pairs of fish are collected from an isolated location. These may be able to spread the genetic pool in the wild when they move around via wet season flooding or even hopping from pool to pool,

KN PRESS - T.R. GRADY

but in the isolation of an aquarium, the only fish may be a single pair which breeds and their young become the sole population for several generations. These fish take on nearly identical appearances over time and at some point, the genetic system fails.

There is only one real answer for this issue - to attempt to obtain different genetic lines to mix into the home population. Even this may not be the solution, particularly when the original importation into the hobby is limited to only one or two pairs. Sometimes new blood from the wild is the only answer to maintain a healthy group of fish.

Velvet or Gold Dust Disease:
Velvet, a dinoflagellate parasite, is also known as the 'Gold Dust' disease, because the fish become covered in what appears to be a fine dust of gold to brown coloration.

The life cycle of *Pascinoodinium is* in three stages. Basically, the parasite, known as a tomonte, rests in the bottom substrate and divides up to 256 times after which it enters a mobile stage (tomite) where it seeks a new host. Interestingly, during the second stage the dinoflagellate will use photosynthesis to grow. The third stage is when it enters the mucous coating of a fish and begins to dissolve cells. In about three days, the parasite detaches from the host and enters the first stage again, splitting into many new adolescents and expanding the infection exponentially.

The earliest symptom of a fish becoming infected is an attempt to scratch itself on ornaments and the bottom substrate. Some hobbyists call this 'flashing' because when the fish scratches, it twists and the sides are often reflected. Next, the fins will become clamped, held close to the body and the host becomes lethargic, often sitting in a corner of the aquarium and gasping due to inflammation of the gills. The fish will ignore food. Obviously by this point, the fish is covered in 'gold dust'. Left

untreated the parasite repeats its life cycle and the fish becomes more and more covered with Velvet. In time the fish will die from necrosis, the dead tissue from the site of the infestation.

Velvet may well be the major cause of a massive die-off of killifish fry and some of the annual species are particularly susceptible to it. Because of the size of the baby fish, it is very difficult to detect the parasite without magnification. It is likely the parasite remains in the peat moss used to incubate the eggs and becomes active once in water again. The addition of one teaspoon per gallon of aquarium salt to the hatching container is highly recommended.

Treatment of Velvet: Velvet is treatable and several commercial products work well. Essentially a combination of higher temperatures in the aquarium (above 80ºF-20ºC) to quicken the life cycle, a medicinal dye such as malachite green or acriflavin and aquarium salt will treat this condition. Commercial products often contain copper sulfate or zinc. Because *Pascinoodinium* uses photosynthesis, some hobbyists believe darkening the aquarium may help. The overall treatment needs to last for up to three weeks in order to affect the long-term life cycle.

Apparently once a fish has had Velvet, it can develop an immunity according to some studies. Healthy fish do fend off infection. Like many of the problems affecting aquarium fish, prevention entails making certain the water quality is high. Generally fish weakened from injury, the wrong water chemistry, rapid temperature changes and sometimes the wrong diet are more susceptible. Velvet can be introduced into the ecosystem with the addition of new aquarium stock.

Swim Bladder Issues:
While not specifically cause by disease, Swim Bladder problems are faced by killie hobbyists on a regular basis and can be frustrating. There are actually two different issues that can be identified: belly-slider

fry and swim bladder damage.

Belly-Slider Fry: In its simplest description, this problem is normally found in annual killifish fry. There are a variety of reasons this occurs, but the primary culprit is thought to be either underdevelopment of the swim bladder or over development. This is caused by either hatching annual eggs before they are fully developed, or hatching them after they have passed the maximum development period.

It is not known if this occurs in nature or is an aquarium-only problem. There are no reports on this in the available literature. If it occurs in the wild, then the fry most likely die within a very short period, either as prey to something else or simply the inability to thrive.

In the home aquaria there are a couple tricks which may help to alleviate this problem. Some breeders like to hatch the eggs in very cool water. This is thought to be a duplication of the first rainfall which can be as low as 40ºF (5ºC). The fry hatch in this cold water and the air already in the swim bladder expands to provide the necessary balance.

Another theory claims highly oxygenated water will make a difference. One study done in the 1960s indicated hatching the eggs in a very heavily aerated jar where the eggs are continually agitated by the airflow does not produce belly-sliders.

While for many years it was thought the fry needed to swim to the surface to take a gulp of air to fill the swim bladder, this is no longer considered a viable hypothesis.

Swim Bladder Problems in Adult Fish:
The other problem comes with adult fish who lose the ability to swim. This can be caused a couple of ways and there is little or nothing that can be done to help the fish, although they may live a reasonably long life and even reproduce.
The first cause is injury to the swim bladder while attempting to reproduce. Most annual species of killifish press against a hard surface (tank bottom or breeding media container sides) while actively spawning. It is possible this can lead to damage.

The second possibility lies in a bacterial infection of the swim bladder that can be caused by heavy bacterial growth in the aquarium. The hobbyist has two options, either euthanize the fish, or allow it to live out whatever life remains.

What are Those Creepy Crawly Things in My Squarium?

Infestations are sometimes serious problems as a wide variety of creatures can invade aquariums and often lead to a decline in the health of any fish. There are any number of creatures hobbyists do not need crawling on the bottom or up the sides of their tanks and in some cases directly attacking the fish.

For hobbyists who harvest wild daphnia or mosquito larvae there is a real possibility dragonfly nymphs or water beetles may also be in the mix. These can and will feed on the fish and must be eliminated. There are other types of invaders which come in on plants, are transferred from the tanks of friends and aquarium stores or in some cases, simply arrive from an unknown source.

The single best way to protect killies is preventative - know exactly what you put in your fish tanks and where it came from. Even this may not prevent an infestation. The following pests are some of the more commonly ones found in home aquarium environments.

Hydra and Aiptasia: One problem every hobbyist faces at some point is *Hydra*. The anemonie *Aiptasia* looks nearly identical but only happens in salt water. There are a few species of killifish that do much better in a salt environment so it is possible that *Aiptasia* could infest a tank. Both can be treated in the same manner.

KN PRESS - T.R. GRADY

Tanks can easily become infested with *Hydra* that can come in attached to plants and sometimes are transfered between tanks on nets. Unless the hobbyist is looking for them, the organisms can be difficult to notice in the earliest stages. Once full-blown they can densely cover whole sections of the glass and anything else that they attach. Since the *Hydra* can retract, very often they are not noticed and only make an appearance when they need to feed.

The *Hydra* is a solitary polyp related to anemones and jellyfish and are members of the phylum *Cnidaria*. In its adult form it appears as a white, pinkish or even orange stalk attached to the aquarium glass or ornaments. Technically the bodies of *Hydra* are translucent and take on the colors of their latest food. There are a number of species of *Hydra* and certain ones have their own color variations. For example, because of a symbiotic algae cell, *Hydra viridis* takes on a green aspect. At the 'top' of the stalk are several tentacles which contain stinging cells called cnidoblasts which release stinging barbs known as nematocysts densely spotted along the tentacles into the prey.

The initial stages of *Hydra* appear as numerous tiny polyps on the side of the aquarium. If baby brine shrimp are fed to the tanks, the *Hydra* in this form can be mistaken for unhatched brine shrimp cysts. They are of similar size. *Hydra* will also retract into tiny sacs when stressed.
Essentially the major concern for hobbyists is *Hydra* will feed on fry if they can catch them. *Hydra* will also attach to eggs and use them as a source of nourishment.

While it has been said *Hydra* don't harm adult killifish, it is possible if a fish comes too close, it could be stung. If the colony becomes large enough, multiple stings will sooner or later affect smaller species of fish and over time could kill them. *Hydra* will easily feed on baby brine shrimp, *daphnia* and *moina* which indi-cates they can certainly catch anything moving in the water column. They can infect live food cultures and are then inadvertently transferred into the fry tanks and adult aquariums.

Treatment: There are a few fish which will eat Hydra, primarily gouramies and mollies. While this can be useful in a community tank, unfortunately killie hobbyists tend to segregate species into small ecosystems and this is not a viable option.

Eradicating these pests has become easier in recent years. The introduction of Flubendazole™ to the hobby has become an important treatment for several ailments. Powdered Flubendazole™ is spread across the surface of the water of the tank or infected live food culture. While it is not easily soluble in water, it still seems to be effective. Also Fluke™ tabs and Clout™ work well. Some hobbyists have claimed a copper penny in the tank will eliminate hydra, but copper is not something a hobbyist really wants to add to the water column. Copper over certain levels can seriously affect fish. Some zinc-based medications are also available on the market. Aquarists should always be a leery of introducing heavy metal products into the system. Many are considered poisons in larger dosages.

The *Hydra* should retract into a small sac and eventually drop off the side of the tank/ornament within a day or two if the treatment is effective. If the *Hydra* remain in their normal extended form, then the treatment has failed and a larger dose or an alternative treatment may need to be considered.

Planaria: On occasion hobbyists will see a small arrow headed creature crawling on the bottom or sides of their tanks. Most likely this is the flat worm known as *Planaria*. These 1/4-inch long creatures are members of the *Platy-helminthes* phylum and the class *Tubellaria*. A simple description of the flatworm is a pointed tail area extends to an arrow-headed top. Two eye spots are visible in the head. They glide along surfaces using cilia (hairlike projections) on the underside of their bodies.

Planaria are often associated with a 'dirty' or overfed aquarium, one where detritus and debris litter the bottom and provide a source of food. *Planarians* feed by extending a long tubular pharynx from its mouth which is located on the underside of the body. They feed upon crustaceans, larvae and small worms by secreting a digestive fluid onto its prey and then sucks off the partially digested food.

They reproduce by two methods, the first being eggs fertilized sexually, the other is asexual where the tail splits off the body and forms a new head. The original *Planarian* will regrow the tail.

Planaria are not dangerous to tropical fish directly, but they may indicate a real need to clean the aquarium. The unhealthy condition of the water is what may cause the fish to become ill or die. A combination of water changes, removal of detritus and general good water quality will lead to the flatworms disappearing. The alternative would be to strip the tank completely, washing everything with chlorine bleach including a complete cleaning of the filter system and resetting it up.

There is a product on the market called "No Planaria™" and it will eliminate the pest and apparently other worm and *Hydra* problems and is a popular dipping treatment for shrimp enthusiasts and used for new plants scheduled to be added to an aquarium. It is available on the Internet by Googling "No Planaria™". However, this medication does not eliminate the root cause of the infestation which is a poorly maintained ecosystem.

Insect Larvae

Tube Worms: Tubes or encasements on the bottom and sometimes extending up the sides of killifish tanks is a more common problem than aquarists realize and often is not recognized as to the potential seriousness the tiny

creatures living in the tubes can bring forth.

From observations, these tubes contain the larvae of a tiny, even delicate species of flying insect and the adult seems able to find its way onto nearly any water surface to reproduce. **Initial stages:** On first glance, the bottom of the tank simply appears to have some waste material gathering in one small area. On closer examination it becomes more obvious that the 'dirt' has been gathered to surround a tiny (less than a quarter inch long) cocoon which contains a minuscule worm that sometimes sticks its head out. Initial examination by eye does not reveal anything, but under stronger magnification, it becomes apparent that this 'worm' has a form of mandibles which are used for feeding.

As days pass, it seems like there are more and more of these encasements and over time, the entire bottom of the tank will be covered in them. At times these 'worms' can be seen actually swimming in the water column if it is carefully examined. The pest reaches a stage where it will migrate to the surface of the water and then transforms into a tiny flying insect. These insects stand on the surface and this is where the reproduction process takes place - mating and laying microscopic eggs which hatch at some later time. The larvae migrate to the bottom.

Tube Worms are a common occurance caused by a tiny flying insect. They reproduce rapidly once established. **Photo: Tom Grady**

KN PRESS - T.R. GRADY

While these larvae do not appear dangerous to adult fish or even killie fry, they present a completely different problem for killie eggs. Apparently they feed by attaching to the egg and then boring into the shell to feed off the albumin and yolk and this leads to the death of the egg. What is very frustrating is the process by which this happens. If a hobbyist has an infestation of whatever this fly is, and uses open containers to incubate the killie eggs, then the flies somehow find the container and lay their eggs. The apparent process only takes a few days before the capsules appear and then soon after the 'worms' begin to attach to eggs if not removed. The best option to deal with the tube worms is to remove all of the eggs in the tray, sterilize it and then place the eggs into fresh incubation water.

Other than keeping a constant watch on the egg containers and changing the water on a very regular basis, there is no known way to eliminate this pest other than by a complete sterilization of the source. Even then, if there are flies in the fish room, they will quickly reestablish the colony. There are no known predators of the 'worms' or fish which appear to feed on them. The longer they are allowed to infest a tank, the worse the situation can become. It has to be assumed the worms feed primarily on protozoans and perhaps even microalgae. These encasements can also be indicative of poor water quality.

Other Bugs: In addition, a number of insects, nymphs and larvae that sometimes get into a tank. Most often these appear when the hobbyist collects live food (*Daphnia*, blood worms or mosquito larvae) from ponds or temporary water sources. Dragonfly and Damselfly Nymphs as well as Water Boatmen are dangerous to fish and fry. These generally have to be removed by hand and destroyed.

Leeches: Leeches are commonly mixed into blackworm cultures and not overly difficult to eliminate, but certainly an uncomfortable creature to unexpectedly find. Generally these leeches do not appear to do any harm to the fish, but that does not mean they cannot. Essentially leeches are blood-sucking parasites, so it is preferable to not have them get into the aquarium.

An easy trick to removing them from blackworm cultures is to empty the blackworms into a long, low tray, preferably plastic. Fill the tray with fresh water. Give them a little time to settle and then look for pink to white leeches on the bottom. These will attach to the bottom of the tray using their suckers. Now gently pour off the blackworms into another container. The leeches will remain attached to the original try. This may take several attempts to completely remove the leeches, but should work. Do this on a daily basis because there may well be eggs and minuscule leeches that poured out with the blackworms and will grow to a size large enough to be obvious. Once attached to the bottom of the tray, you can eliminate them with any number of 'poisons' including boiling water or chlorine bleach and then send them down the drain.

Some people believe these 'leeches' are actually flatworms that feed on the blackworms and nothing else.

Snails: There are no invaders of the aquarium that are as much reviled, yet desired as snails.

The main complaints are a combination of overpopulation and the destruction of plants. Yet, some snail species provide a real benefit by eating detritus as well as stirring up the gravel to relieve the bottom of potential anaerobic pockets of toxic substances. Snails are also essential to well maintained *Daphnia* and some infusoria cultures. Dead snails and trumpet snails appearing during the daylight can also serve as an indicator of poor water quality.

The remaining shell of snails that have died takes a lion time to break down into its component minerals and should be removed.

Still many aquarists would prefer not to have them in their ecosystems and seek removal.

While there is no 'great' method for eliminating snails from the environment, perhaps their willingness to feed on leftover food and the carcasses of dead creatures (live foods and fish) provide a real benefit to the tanks.

There are some aquarists who have reported certain snails will take a killifish egg, roll it over in its mouth to clean it and then release it back into the mop. They do not appear to eat healthy, fertilized killifish eggs, but may feed on dead eggs. There is some question as to the validity of this, but it is mentioned for no other reason than to make the reader aware.

Elimination of Snails

- **Chemical additives** containing copper will kill the snails. This is probably the last option a hobbyist should use. Dead snails decompose no differently than any other denizen of the tank and that can leaf to ammonia problems.
- **Predators.** Several species of fish feed on snails including Clown and Yoyo Loaches, puffers and goldfish. The question then becomes one of value. If you are breeding killies, does the hobbyist want to place these fish into a community situation?
- **Hand Picking.** Certainly time consuming and there is no guarantee the tank will be free of eggs or tiny snails.
- **Sterilization.** Stripping down the tank. Who likes completely emptying a tank and starting it again from scratch?
- **Food Traps.** While not a complete solution, snails seek food like any other animal. Plastic container tops floating on the surface seem to

draw the snails and can be easily removed with many snails attached. Sinking plates or weighted clips can be used with some vegetative matter attached. Place this in the aquarium at night and snails will gather to feed. Simply remove the trap and repeat the next night.

KN PRESS - T.R. GRADY

Reproduction: Successful Breeding Techniques

While killies are among the most beautiful fish in the aquarium hobby, the longterm draw for most aquarists is a compulsion to successfully reproduce the fish. No other group of fish offers as much diversity or challenge for hobbyists interested in breeding them.

In some case, there are a few species it seems only need water to induce them breed. Numerous others require a real effort to discover the tricks and care necessary to bring them into spawning behavior and demand a specific environment or they simply never reproduce.

Killifish designated as annuals require a bottom or bowl filled with some sort of 'dirt' in which to 'plant' their eggs. That substrate normally requires a dry period for the eggs to successfully incubate.

Mop-spawning species require a substitute for plants in which to release their eggs. Some of these killies desire cooler temperatures or a specific pH or hardness for successful reproduction.

In addition, there are some species which will switch between plant spawning behavior or, under certain circumstances, choose the substrate in which to leave eggs.

This section is an overview of several spawning techniques as well as suggestions about hatching and raising killifish fry with a short discussion about the properties of killifish eggs and the resting stages known as diapause. Sex ratios will also be discussed.

Killifish Eggs

Killifish eggs are unique. They are hard shelled and can be handled in a hobbyist's fingers. This can lead to any number of ways to handle and incubate them.

Unlike most species of tropical fish, most killies need at least two weeks and in many cases several months to develop while cichlids and tetras take a few days and live-bearers drop free swimming babies. Only Rainbow fish and their cousins the Blue-eyes offer similar habits to some of the killies.

Sex Ratios

Killifish keepers have been known to rant and rave over the problem of sex ratios produced by their fish. Sometimes every fry turns into either all male or all female and no one seems to have a good answer as to what has caused this problem. It has led to the loss of a species within the hobby. Theories range from water hardness/pH/temperature variables to hormone issues. Strangely, there are numerous reports of only two fry reaching maturity and

they turn out to be a pair. There are as many suggested theories for this also.

For the skewed sex ratios, there can be some simple explanations which include disease killing weaker fry, predation of rapidly growing fish and larger males killing their siblings. None of these really fully explain why a tank becomes only a single sex.

A breeder should expect a 50/50 ratio of male to female, or at least that is most believe. However nature doesn't always see things the same way. Actually two major factors figure into sex ratios of killifish: genotypic where the chromosomes determine the sexual orientation as well as environmental- variable in chemistry and temperature.

In a perfect situation, genotypic selection should be 1:1, an equal distribution of chro-

Aphyosemion australe Gold in the process of laying eggs in a mop. **Photo: Mike Jacobs**

mosomes since females provide only female (X) and males provide the male (Y). It takes two chromosomes to form a viable fry which leads to a random ratio consideration of 50/50. Admittedly, this can go bad through mutation or an incomplete transfer of genetic information. The sequence of chemical triggers can be affected during early growth leading to variations.

Environmentally, it has been proved in cold blooded animals, in particular reptiles, that temperature plays a large part in sexual determination. The closer to the ideal incubation temperature is in the nests, the closer to 50/50 the sex ratio becomes.

Influencing Gender in Killifish:

For some time, it has been known that gender can be influenced by the environmental factors of temperature, pH and hardness, age and inter-estingly some species where simply the death of a male will lead a female to change gender (*Labroides dimidiatus - a marine cleaner fish). A study on dwarf cichlids some years ago indicated the combination of two or more of the environmental factors could lead to specific gender selections. For example low pH plus high hardness would produce predominantly males, while high pH and low hardness allowed for the females to dominate.

No specific scientific studies have been completed on this topic, but a hobbyist may want to experiment a little. It would of some value to see published results.

---

* Reference: Aquarium Science Male or Female? Gender Deterimination in Fish by Donna Recktenwalt TFH Magazine February 2007.

Annual Killifish: Peat Spawners

In both Africa and South America there are large groups of Cyprinodonts that are considered "Annual Killifish". Among those many groups are the better known *Nothobranchius*, *Austrolebias* and *Simpsonichthys* species.

Too often the term annual is misunderstood to mean the fish are short-lived. While this can be true in some cases, the real definition of annual refers to the natural life cycle of the environment from which the fish originate. In large portions of both continents, cyclical wet and dry seasons effect the environment and the lack of rain leads to periods where water in the pools and ponds evaporates and the land becomes cracked and dry.

Several groups of killifish have adapted to this cycle by laying their eggs in the substrate. The waters dry out and the eggs rest in wait for the next rains to refill the reservoirs. During this time the eggs incubate and then enter resting stages known as diapause. Diapause is interesting because there are several stages and not every egg will follow the same development process. Instead some eggs will remain in the resting stasis well past the time others are ready to hatch. This occurs to allow for times when rains fall in an unpredictable pattern and can fill the habitat temporarily. Some eggs simply are not ready to hatch and remain dormant until the next rainfall or even the one

Nothobranchius ugandensis   **Photo: Tom Grady**

following that. This allows for the continuation of the species.

In Malawi recently an extended drought lasted over two years, but when a few locations were checked once rain fell again, Nothos were found.

Part of the intrigue of successfully reproducing these fish is the process the eggs go through. Nature has found a way to ensure the survival of the species under the most difficult of conditions even when a dry period lasts years instead of months.

In the home aquaria, the breeding and incubation is a process that is replicated by the use of peat moss and coconut coir among other spawning media for breeding and then processing those substrates containing eggs. The substrate is allowed to dry to a certain consistency, placed into plastic bags and set aside in an incubator for a period of incubation. Once

The series of photos avove depict a six to eight month period demonstrating the changes in environment in a typical Nothobranchius habitat.

-

that dry period is over, the entire contents of the package are placed in water and the eggs should hatch. That may be a simple description and in practice, it is a bit more complicated.

To successfully breed annuals, there are several techniques, all based on a single concept - the use of some form of spawning media to replicate the bottom of pongs the killies inhabit and reproduce.

The simplest and most basic method is to introduce a container filled with processed peat moss (or coconut coir) into the tank and allow

Pyrex glass bowls do not need additional weight to hold them to the bottom of a tank.   Photo: Tom Grady

the fish to use it over a week or two (or longer) to breed.

Those containers range from plastic containers (with a layer of marbles to weigh them down) to heavy glass bowls or glass goldfish bowls. Some breeders like to use a cover on the containers with a hole cut into the cover large enough for the fish to easily enter. The idea is to keep as much of the spawning media inside the container as possible. The fish will discover the opening.

Once a breeding container is introduced into the aquarium, the males will find the media, stake a claim to it for a temporary period and attempt to entice females to join him. When the female agrees a few eggs are released into the folded anal fin of the male and fertilized. With a flip of the body and fin, the eggs are then buried in the media. This is an ongoing

process and well conditioned pairs or groups can produce hundreds of eggs over a week long period.

Preparing Peat Moss and Coconut Coir

Whether peat moss or coconut coir is chosen, the substrate needs to be saturated with water, washed and processed.

There are a number of sources and types of peat moss. Jiffy™ Peat Pellets are available in most garden shops, but the hobbyist needs to be careful not to purchase the variety containing fertilizer. Peat pellets are already sterile. They should be placed in a bucket of water and allowed to soak. In about a day the peat should settle to the bottom.

The alternative is the use of 100% sphagnum moss without any chemicals or fertilizers. Sphagnum can be bought in 3-cubic foot bales and Canadian sphagnum is common in most

garden centers.

If sphagnum moss is chosen, it needs to go through a couple of important steps before being placed into holding containers. First, the sphagnum must be run through a blender to reduce the size of each strand and then it needs to be boiled to sterilize and kill any potential pests. The final step is the rinse to remove the finest particles which remain suspended in the tank water for long periods of time and might clog up filtration systems. The use of a fine mesh net under a running faucet does this quite well. What is wanted are the course fibers.

Once the peat is washed and the hobbyist satisfied with the result, it is time to put it into the breeding container. Make certain it is at least two inches deep when moist and then add water to the bowl until it is full.

Place the top on or if you do not want a per-

manent top, fit a top with a hole in it. This top should be loose enough so once sitting on the bottom of the tank it can be easily removed. As water enters the container, expect some of the peat or coir to escape. It will not harm anything. If the open container is desired, then slowly and carefully remove the covering and allow the peat/coir to settle.

It will not be long before the fish begin to examine the bowl and finally spawn. Once the fish begin to use the peat container, expect the fibers to get spread around the tank. This is one good reason not to have gravel covered bottoms. The permanently covered containers are less likely to create this problem.

Storing Annual Eggs

Okay, now that the annuals have spawned, what is the next step?

It's time to remove the peat moss or coir and prepared it for dry storage.

First remove the spawning container from the tank carefully and set it to one side. Next

paper towels (newspaper, bathroom brown towels, etc.), into a thin layer and then cover it with more paper towels. The use of trays can help hold the spawning material in one place and absorb more of the dampness. Prior to covering the peat, it can be searched by eye or a magnifying glass to see if eggs are visible. Set

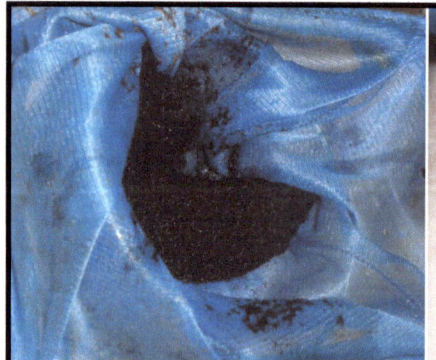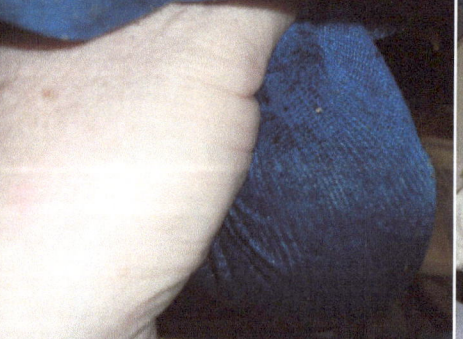

Peat in the net is squeezed until no water comes out and then placed on paper towels to finish frying

carefully siphon any extra fibers that have exited from the container into a small pail. Peat often flies out of the container when the fish become rambunctious during spawning. Pour the water from the breeding container and then the pail through a fine mesh net to remove the peat or coir fibers. Gently squeeze the net to remove the remaining water from the media until it no longer easily produces any liquid and then distribute the media over

this aside for 12 to 24 hours to continue the drying process.

A popular comparison of how dry the peat needs to be for the final storage could be described as similar to tobacco dampness and consistency. If the very top edges of the peat or coir appears light brown or dry, then chances are the media is ready for storage.

To package the peat for incubation, simply fold the paper towels in half, slip them inside a 1-quart size plastic storage/freezer bag and shake all of the peat that remains. Remove the towel and make certain all the peat and eggs have been removed. The baggie can be sealed either tight to the peat or loosely and placed into an incubator for a predetermined time or

Store the dried peat moss in plastic bags and place into an incubator for a designated time period.

until the eggs are examined and demonstrate readiness to hatch.

Incubation of Annual Eggs

Incubation of annual eggs varies greatly from species to species as well as to the temperature in the incubator. The shorter the dry season in nature the less time it takes for the eggs to reach complete development. Equally so, the higher the temperature in the incubator, the shorter the time period for the eggs to be ready to hatch. In the home environment, many hobbyists try to find a stable temperature in order to build a specific time frame for the eggs to hatch. There is some disagreement on the best options and temperatures.

Many hobbyists build an incubator that can maintain a consistent temperature of 75ºF (24ºC) and base the incubation period on that. Some others prefer to incubate eggs at a high-

er temperature in order to shorten the time before hatching. Humidity may also play a part in the process.

One experienced breeder theorizes that nature provides alternating periods of warmth and coolness (day and night) in a 24-hour time frame. The suggestion is that eggs expand and contract under those conditions and this affects development. This concept recommends a timer to turn the heater and any fans on and off for 12-hour durations.

There are also as many incubators as there are hobbyists. The simplest may well be a styrofoam box placed near the ceiling of a fish room. Chances are there will be variations in temperature with this system based on time of day, the heating of the room and the time of year. However it works perfectly well. Others use a styro or build a box (below) and place a container of water with a small electric water heater inside. This is adjustable and will maintain a constant temperature inside the container, but the potential for cheap heaters to break and cook the box exists. This also method also needs to check the water levels to replace evaporated water in the jar.

<---Water Container with heater

Dr. Dan Nielsen's incubator

KN PRESS - T.R. GRADY

A similar option is to purchase a chicken egg incubator.

If a breeder is so inclined and breeds a large number of species, the idea of a permanent incubator with temperature controls can be considered. While there are a number of designs, one which works very well was designed by Dr. Brian Watters. This system uses several shelves made with screen to hold bags of eggs, light bulbs attached to a thermostat to turn on and off to maintain a constant temperature and a small fan to maintain movement of the warmed air between the different levels. The majority of the incubator is made with plywood and styrofoam.

Alternatively, if a larger enclosed (a closet or metal self-contained storage unit) area is used, there are numerous open metal (see photos) and plastic bins that can be used to hold the bags of eggs and attached to the walls, shelves or even purchased as a free-standing rack. A small space heater with a built in thermostat and fan is placed on the floor and set for whatever temperature is desired. This will turn on and off as needed.

Essentially the options are endless because the only real requirement is to create an environment conducive to incubating the eggs.

When are the Eggs Ready to Hatch?

Annual eggs need time to incubate and there are a couple of thoughts on how to determine the time period the eggs need to remain dry. For many hobbyists, specific conditions - the temperature of the incubator and the use of well established charts based on personal experience and/or the experience of other breeders provide a valid chart to wait for each species.

Hobbyists can consider three months as the minimum time frame for most annuals, but many species from both Africa and South America need much longer. Species such as *Nothobranchius guentheri* (Zanzibar) and *N.*

korthausae (Tanzania) can be hatched at two-three months when incubated at 76ºF (24.5ºC) and *Austrolebias nigrippinis* (Argentina) is about the same, while *N. rachovii* (Mozambique) and *Meg. wolterstorfii* (Argentina/Uruguay) may take up to six months or longer.

Higher incubation temperatures shorten the time necessary for full development. One concern with incubation periods is healthy development of the eggs. Belly-sliders (fry which cannot swim due to damage to the swim bladder.) sometimes appear when eggs are hatched that have not completed development, or in some cases have remained in a dry state for too long.

An alternative point of view is to examine the eggs at certain intervals of time (2 months, 4 months etc.) or attempt to determine the length of the dry season where the species resides in nature before examining the eggs. The best way to look at the eggs is the use of a dissecting microscope (binocular microscope). Look for a gold rim around the eyes inside the egg. It is also possible to see a beating heart when light strikes the egg. If the gold rim is visible, then it is time for the eggs to be placed into hatching water. Most should hatch within 24 hours, but a few may take a couple of days. If the eggs do not hatch within 48 hours, the peat should be re-dried and placed back into the incubator for an additional thirty days.

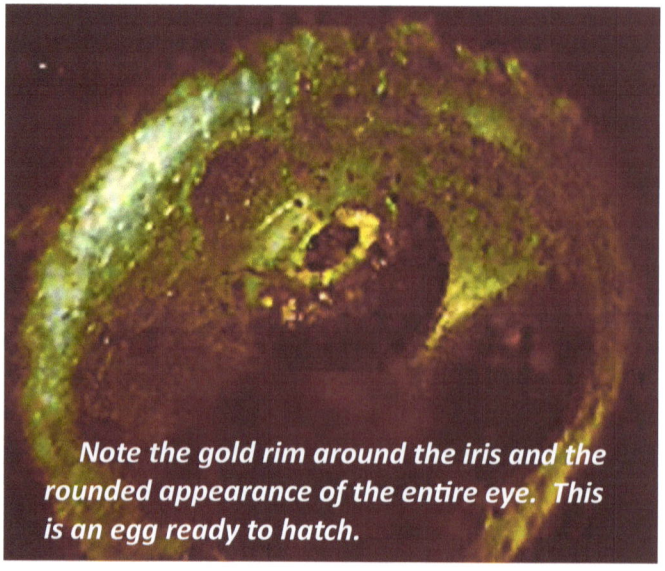

Note the gold rim around the iris and the rounded appearance of the entire eye. This is an egg ready to hatch.

Mop-Spawners: The Plant and Mop Breeders

A vast number of killifish spawn in plants in the wild and killifish hobbyists have found several ways to provide an alternative place to lay eggs in the home aquaria as well as several concepts to store and incubate the eggs until they hatch. In reality, most eggs are best incubated in containers of water. There are many killies for which the use of damp/dry storage works quite well.

There are literally hundreds of species that can be bred by using spawning mops made from acrylic or nylon yarn. The eggs are picked from the mops and paced into containers to be incubated for two-to-three weeks.

The fish use these mops in place of plants

Making a simple Spawning Mop:

There are numerous ways to create mops, some quite elaborate. The following is a basic mop designed for quick use.

Materials:
- Acrylic or nylon yarn. The use of other types of yarn is dangerous because chemicals/dyes can be released into the water.
- Floater (cork/styrofoam, etc.)
- A book of about 8 to 12 inches in height or width.
- Hot to boiling water
- Scissors or a sharp knife.

Wrap the yarn around the book to the length desired up to 50 times (which creates a 100-strand mop). The actual number of strands is not important and only what the aquarist determines. Carefully slide one end of the yarn from the book and slip a length of the same yarn through the circle and tie it off.

The extra yarn can then be wrapped around the floater and tied off. Stretch the yarn out and using a pair of scissors or knife, cut the yarn directly opposite the floater. (An alterna-

and the hard shelled eggs can be removed by hand and placed in storage. Color of the mops may not be important, but the use of a variety of greens and browns seems appropriate. Some species tend to prefer to spawn near the bottom of mops while others will place the eggs as high into the yarn as possible and just beneath or even on the floaters. Some members of the Rivulus groups will lay the eggs on the edge or even out of the water.

There are some species of killifish which prefer to lay their eggs in crevices, but the same basic mop making materials and techniques can be used with some minor alterations in the end result.

KN PRESS - T.R. GRADY

tive is to place the floater through the circle of yarn and tie it off beneath the floater.)

The completed mops should be placed in hot or boiling water for about 30 minutes in order to remove any extra dye or other contaminants that may remain. This also helps saturate the mop with water so it will hang in the water column. This is a 'floating' mop that extends from the surface of the water to its full length.

To have a mop that rests on the bottom of the tank, simply do not add a floater, but it is made the identical way. Both ends can be tied off and for crevice spawners, a tie every inch to 2 inches creates the necessary tightness in which the fish force their eggs. For crevice spawners, a float can be attached at both ends or not be used at all.

Planted Tank Method

Many hobbyists prefer not to pick eggs on a daily/weekly basis and they often set up a heavily planted tank, generally using Water Sprite, *Salvinia* and/or a mix of other floating plants. The killies will breed in these tanks and a percentage of the fry will survive to adult size. In fact, many species of killies are not cannibalistic. Certainly this simplifies the process, but it is also likely to only produce a few fry at a time.

A few aquarists will look for fry swimming at the surface of the tanks and remove them to fry-rearing containers. These often become very nice community tanks that maintain a restricted, but stable population for many years. As fish are removed for shows, sales or to pass

on to other hobbyists, young ones often appear to replace them.

Peat Sphagnum Method

This is a variation of the Planted Tank technique that works very nicely for a number of harder to breed species. While the numbers of young fish that are produced may not approach the levels of those from which the eggs are picked, this method is good for almost any killie. It works very well for members of the *Diapteron* group as well as other sensitive West African species of *Aphyosemion* including members of the *Cameronense* group, *A. herzogi*, *A. joergenscheeli* and *A. wachtersi*.

Essentially, this method employs the use of coarse long fiber **Sphagnum Peat Moss** that is

not blenderized. The fish breed in the fibers, the eggs hatch in that tank and the fry can hide from the potentially cannibalistic parents. When fry start to appear, the parents can be moved to another tank set up the same way and the fry raised in the original tank.

The major drawback is the effort necessary to clean the tank. Because peat moss can affect the water by lowering the pH, the hobbyist needs to be aware of that potential as well as very fine particles of the peat which can seriously cloud the tank and make it hard to catch the fish if they need to be removed. Obviously if the fry are in the peat, this too is a problem if the hobbyist wants to separate them before

> Some breeders advocate removal of the parents to allow the fry to grow.

they are large enough to be easily seen and caught.

Weekly or bi-weekly water changes are encouraged and at that time the fine particles can be carefully siphoned from the bottom and fry may be removed from the breeding tank at the same time if large enough. Every killie-keeper develops their own way to accomplish this. In many cases, the fry will grow to maturity in the breeding tank.

Egg Trays & Containers

All sorts of techniques are used by killie-keepers to hold and incubate the mop-spawners eggs until they hatch. Among the popular choices are small plastic food containers and clear plastic boxes used for screws, bolts and other small construction needs. The pictures below demonstratetwo popular options. The choice of container may be dependent upon the space needs of the hobbyist.

No matter what is used to hold the eggs, it is absolutely necessary to keep these containers as clean as possible.

Water changes in the egg containers is valuable. Eggs actually breathe and give off respiratory wastes. Changing the water regularly cuts down on these wastes.

Most species hatch in eighteen days to three weeks, but some can go as long as six to eight weeks in water incubation. A good rule of thumb is to consider most West African species and *Rivulus* species run around three weeks while *Fundulopanchax* tend to take longer. In fact, most members of the *Fundulopanchax* group can be stored on very damp peat moss for several weeks and then, when eyed-up, induced to hatch by adding water.

This technique uses peat moss placed in a covered plastic container and the eggs individually placed on top of the peat. They can be watched for development this way. Generally it takes six to eight weeks for this process, but the temperature of incubation also plays a part. The warmer the containers are kept, the shorter the incubation period.

Protecting the Eggs During Incubation

The initial care of killie eggs takes on one very important aspect from the moment they are picked - prevention of fungus which can kill the eggs and can spread rapidly to infect viable eggs when given the opportunity.

Recognizing bad eggs: For the most part, by picking eggs by hand, the hobbyist can initially determine very quickly a high percentage of unfertilized eggs by finger rolling them around with a small amount of pressure. Good eggs are not harmed by this, but unfertilized ones will be squished. Also any eggs that are not clear, but instead are white are also bad. The white and unfertilized eggs are the perfect breeding ground for fungus to become attached.

Protecting Good Eggs: A combination of dyes, fungal inhibitors and tap water provide a good

KN PRESS - T.R. GRADY

barrier for the eggs against fungus. Despite the chlorination of tap water, the chemical actually has some benefit for killie eggs. The chlorine generally dissipates from the water within a few hours and does kill most potential invaders. When a fungal inhibitor and a dye is added to the mixture, it only increases the protection of the eggs.

The use of the dye methylene blue also is useful in two respects. Since eggs 'breathe', the

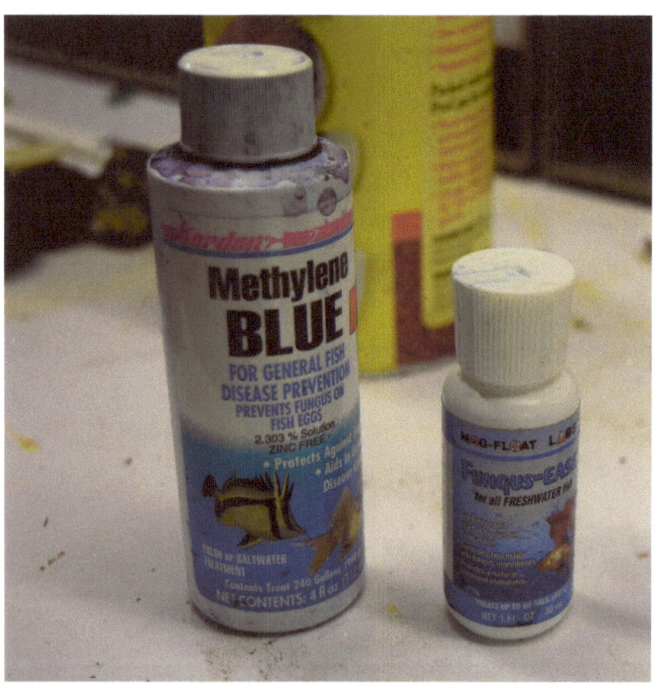

dye will be drawn inside bad eggs and color them blue. Those eggs need to be removed from the incubation container before they begin to fungus. Methylene blue also darkens the water and helps protect the eggs from too much light. Malachite green is generally considered a viable substitute.

The use of acriflavin, while often suggested by the commercial interests, to protect eggs, it is not recommended. While it has not been specifically studied, there are some aquarists who believe this dye can alter the DNA of the eggs and potentially cause genetic defects. This is only rumor, but it probably reasonable to be cautious. However, acriflavin is considered a poison by the Centers for Disease Control in the USA.

To inhibit the spread of fungus to good eggs, the use of a fungus medication is recommended. There are a variety of anti-fungal medications on the market and most work fine. Many hobbyists actually make an incubation water mix of a fungus med and methylene blue. A one gallon container is filled with tap water and then up to five drops (or recommendation on the bottle) of the fungus inhibitor is used and enough of the dye is added to color the water medium blue. This water is then used in the incubation containers and should be changed every couple of days.

The mixture provides ongoing protection against any fungus becoming established on good eggs even when some bad ones do show the thready cotton-ball growth. For obvious reasons the bad eggs should be immediately removed as soon as they are seen.

The other important consideration is light. Some eggs are sensitive to light and too much illumination can affect the development of the eggs. It is better to incubate eggs in dim light or perhaps even better in a dark location.

Raising Killifish Fry

Clean water and water changes are a necessary evil that must be scrupulously observed for optimum growth of young killies. While there are many ways to accomplish this, it always involves enlarging the water volume of the fry trays to tanks to larger containers. It does not matter if the fry are hatched out in small egg containers or if they come from a bag of dried peat moss. The overall process remains the same.

same tank over and over, the idea is to double the volume by filling smaller tanks and then transferring everything to larger ones. This is done daily until a 10-gallon or larger tank is filled. By that point the juvenile fish should be large enough to be easily maintained and in some cases may begin sexing out. However, this does not preclude occasionally siphoning out detritus and left-over food.

Step 1 is to remove newly hatched fry from egg containers. Use of an eyedropper with a hole

Velvet is the scourge of killifish fry. It is very hard to detect on babies without magnification, but it is probably the number one killer of young killifish. If the fry seem listless and close examination reveals clamped fins, then it is very likely Velvet is the culprit. Careful examination with a magnifying glass will reveal the telltale gold dusting of the parasite on the body. Even using commercial medications is no guarantee of saving the fry. Sometimes the treatment works, but other times the entire hatch will perish. The one preventative measure which helps with most of the species that tolerate harder water is aquarium salt. The use of a teaspoon to a tablespoon per gallon of water may be used from the start and it is rare for velvet to make an appearance. The additional benefit of the salt is brine shrimp nauplii survive far longer.

The first container used to raise fry should be a shallow tray or a small tank with only about two inches of water. Floating plants such as water sprite or foxtail may be added to provide security for the fry. Snails and *Daphnia* are highly recommended, but not required. Snails eat the left over food (and any dead fry) and *Daphnia* help control bacterial growth and ultimately provide an additional food source. The first stage container should have some clumps of hair or matted algae or even some green water in it. There will be many microscopic creatures in the vegetation that allow the fry to begin feeding immediately. In addition, newly released *Daphnia* are small enough for the fry to feed upon.

Step by step Process: This concept provides regular water changes in the rearing tanks as fresh water doubles and then redoubles the volume over a period of six to eight weeks. Instead of removing water and replacing it in the

large enough for the fry to be caught without harm or a spoon to scoop them up are the two most common ways. The individual fry is transferred into a low tray - generally no more than two inches deep for the first few days and the water should be aged, perhaps even taken from established tanks. Water changes consist of removing detritus and left-over food with a baster. The removed water should be replaced with fresh.

If a bag of peat is hatched, a larger container (1-gallon plastic tank or larger) should be used.

KN PRESS - T.R. GRADY

There should be 3 or more inches of water in it.

Snails and daphnia should be introduced into the fry tray. Snails eat the dead and left over food and and daphnia help control bacterial growth and ultimately provide an additional food source.

For hatching fry from peat, there are some aquarists who believe cold water should be used to initiate the hatching process. It is possible the colder water may be of some benefit in helping the swim bladders to fill and minimize the possibility of belly-sliders. A temperature of 60º-65ºF (16º-18ºC) is suggested. The theory behind the use of cold water is based on the temperature of rainfall is generally much colder than the air and ground temperatures. As the water warms around the fry, the air held in the swim bladder expands.

An older theory has the fry go to the surface of the water to take in a gulp of air to fill the swim badder. For this reason many breeders do not add any more than three inches of water to the hatching container. There is no doubt annual fry do head for the surface upon hatching, but it may well be a survival mechanism to

(Note: Whether or not either of these concepts have any basis in science has never been tested to the author's knowledge.)

seek a hiding place.

Feeding: Baby Brine Shrimp, vinegar eels and microworms should immediately be fed in small amounts a few times daily. If the fry are very small, it would not hurt to start them with *paramecia*/infusoria if the culture is available. Hopefully the vegetative matter will contain enough microscopic life if no infusoria is available. Most killies have no problems starting on baby brine shrimp however.

Step 2 requires the contents and fish in the fry tray be integrated into a larger container. A two-1/2 gallon tank is a good choice if available, but a 5-gallon tank is fine. The water depth should be at least the same as the original fry tray. Maintaining the same depth of the water will more than double the volume in the tray and is the latest water change. Over the next several days, continue to add fresh water every 24-hours to the tank until it is full. Siphon out any uneaten food (normally a pink cotton ball of dead BBS in a corner.)

Step 3 moves the fish to a larger tank designed to be a permanent homeand is the final step. A ten-gallon tank usually allows enough room to grow out the fish to maturity. At four to six weeks, most killies are large enough to begin feeding on Grindal worms and shortly thereafter frozen adult brine shrimp and white worms.

Grow out 10-gallon tank for *Fp. gardneri* Innidere with a mop to give the species a chance to hide.

The Animals: An Overview of Killifish

Aphyosemion striatum Lamberene
Photo: Richard Pierce

With well over 1,000 species of killifish currently described and new species being added on a nearly monthly basis in addition to the addition of hundreds of isolated populations of many species, the first volume of **The Killifish Encyclopedia** will only focus on the overview of the various groups of Cyprinodonts based on which continent they originate. Following volumes of the Compendium will explore the majority of the groups on a species by species basis.

The adaptability and diversity of these animals is amazing. Killies survive in the the arid deserts of the American Southwest and the high mountain lakes of the South American Andes. They have adapted to periodic complete drying of the pools and ponds in East Africa and to the rapidly changing conditions of Amazon rainforests as well as the salinity of the oceans. One species even feeds on a tiny sunlit algae-covered shelf within the depths of a cavern.

The African continent may have the most species and greatest diversity, but South America and North America are not far behind. Every continent has incredibly beautiful fish along with ugly little brown fish with unique habits. For breeders the challenge of success varies with the determination of the hobbyist in understanding which niche each species holds in nature and how to reproduce a simulated environment to replicate that niche. The idea one that of the largest killifish (*Lamprichthys tanganicus*) succeeds by inserting their eggs in tight rock crevices in Lake Tanganyika. This location is home to predatory cichlids who rule the waters. It forces the aquarist to develop new methods to provide a working environment.

KN PRESS - T.R. GRADY

Taxonomy: One of the frustrating issues to plague the hobby is how taxonomists operate. In some cases these specialists have used geographic separation to compartmentalize groups of killifish while others based thier efforts on bone structure and/or other taxonomic issues.

In the past decade, genetic studies have confirmed some differences, but in others, DNA has simply lead to disagreement or confirmation of earlier identifications and relationships. The names of killifish are in constant flux, but that is true for nearly every other type of tropical fish from live-bearers to cichlids.

Unfortunately taxonomy is an issue that not only affects the hobby, but in a larger sense efffects all fields of zoology.

The Encyclopedia's listing of killifish is not meant to be a definitive resource to the names or placement of killifish. Instead it is meant to be a reference designed with the hobbyist in mind. The identification at the genus level is more in line with how the fish are viewed by both North American and European hobby communities with some attempt of avoidance of strongly held differences of opinion in any group of aquarists. The specific volumes will be focused on the species level.

For example, most intermediate hobbyists know that *Roloffia* is a dead term scientifically, but it is still in common use in hobby circles, particularly by European members. For the purposes of **The Encyclopedia**, *Roloffia* is a viable way to identify the groups *Archiaphyosemion*, *Callopanchax* and *Scriptaphyosemion*. While it is not scientifically accurate, by this usage, all but the newest hobbyists know which group of fish are being discussed.

Aphyosemion ocellatum   **Photo: Richard Pierce**

The Killifish of Africa

African Killifish can be divided into several groups which cover nearly the entire continent from Sub-Saharan deserts to the tropical rainforests of the Congo River basin and everything in-between. In a sense, Africa can be divided between East and West Africa for where certain species exist, but it is more complicated that that. The real separation in species comes with the environment, fish that reside in savannas vs rainforests, temporary waters vs broad rivers and even geographical schisms such as the Great Rift Valley that divides parts of the continent from itself.

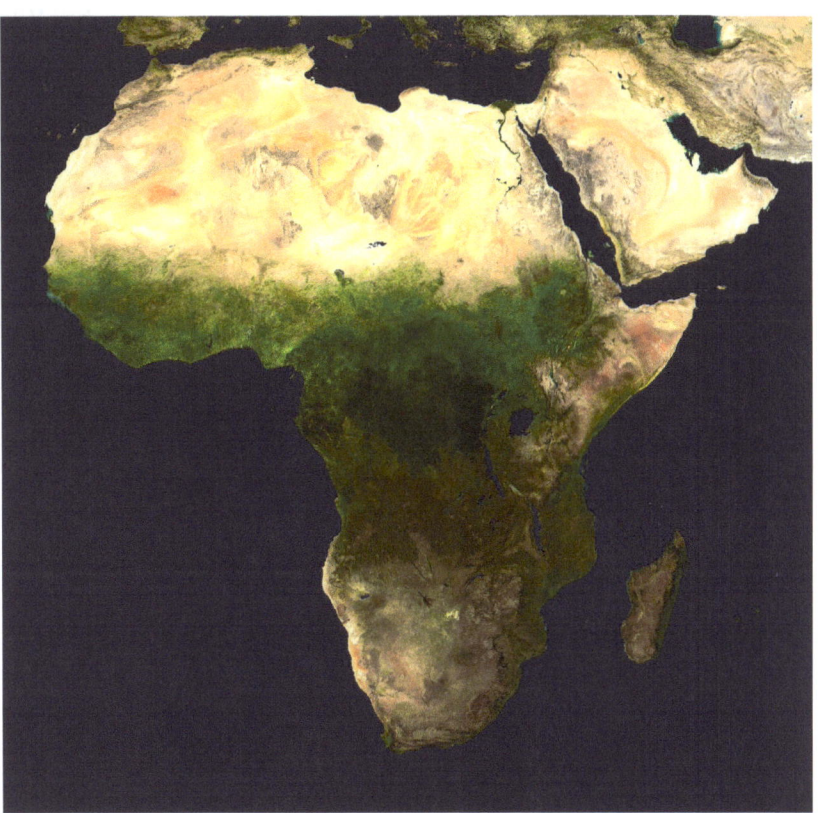

In East Africa, the vast majority of annual killies are related to *Nothobranchius*, but that genus extends further to the east and even reaches the Atlantic ocean in a narrow band across north-central Africa from Somalia into Gambia. A careful study of a map of Africa reveals *Nothobranchius* follow the climate landscape where savannas exist.

In West Africa the majority of killifish are either *Aphyosemion* or *Fundulopanchax* or related to one or the other. These fish live in forested regions and rainforests for the most part and most species are found primarily in brooks and streams. The diversity of these fish is stunning in color. As each river basin is explored, new species and even groupings of these fish are constantly being uncovered.

But there are also the Lampeyes which can be found in nearly every bio-type across Africa. These killies are often found schooling in rivers, streams and other flowing waters, but can be caught in the same locations as the annual and non-annual species.

Add the *Epiplatys* group to the rainforests and the real diversity becomes obvious. Each and every species has found a niche in nature to exploit. One of evolution's choices included predation. It is fascinating to observe the habitat of some Nothos and find small species such as *N. janpapi* may be the primary food for the larger predatory *Nothobranchius* (*N. ocellatus*).

This section on Africa is divided into six regional groupings primarily focused on East and West Africa, but there is some crossover. *Epiplatys* are from West Africa, but also found in parts of North Africa as are some species of the *Aphanius* group which is covered in more detail in the Euro-Asiatic section.

Then there are the unique *Pachypanchax* of Madagascar, a group separated on that island and another single location on the Seychelles in the Indian Ocean. Much like their animal counterparts, the Lemurs, these killies developed in isolation and are different from their mainland relatives.

KN PRESS - T.R. GRADY

Aphyosemion: Rainforest Gems

The Genus *Aphyosemion* is split into a number of groups that have taken on a life of their own, in some cases as subgenra and in others a hobby designation based on certain characteristics or even a historical meaning. In nearly every case, these species all originate in West Africa and in primarily rainforest regions. Taxonomists have used geographic separation to compartmentalize groups while others are based on bone structure and/or other taxonomic issues. In some cases genetic studies have confirmed the differences, but in others, DNA has simply lead to disagreement or confirmation of earlier identifications and relationships. Unfortunately this is an issue that runs throughout the scientific community and affects the hobby, not just with killifish, but nearly every other type of tropical fish group.

That being noted, this overall genus *Aphyosemion* comprises of a number of sub-genus groups, but nearly all are identified as *Aphyosemion* in the hobby. There is no doubt that the groupings are valid scientifically. They include *Aphyosemion* Myers 1924, *Chromaphyosemion* Radda 1971, *Diapteron* Huber & Seegers 1977, *Episemion* Radda & Purzl 1987, *Iconisemion* Huber 2013, *Kathetys* Huber 1977, *Mesoaphyosemion* Radda 1997, *Raddaella* Huber 1977, and *Scheelsemion* Huber 2013. These groups will covered in depth in a future volume of **The Encyclopedia.**

Because many species are found in rainforest regions, the temperature of the water can come into play. Temperatures below 70 ºF (22 ºC) may be required for *Diapterons* and *Mesoaphyosemions*. Another point to consider

***Chromaphyosemion poliaki* Dark CI-01**

Photo: Richard Pierce

with these groups is both pH and hardness can come into play. For the most part, they prefer softer water and a slightly acidic pH. While many species of the Aphyosemion groups will tolerate a fair broad range, for the best reproduction, it is a good idea to understand where each is found in the wild.

Remember this is a generallity and every species has its own requirements.

All of the members of *Aphyosemion* are top and middle spawning killifish. In the wild, the vast majority of *Aphyosemions* are found in the thick plant growth along the edges of the waterways. The use of spawning mops is the primary way they are bred in the hobby and most lay their eggs in the middle or upper areas of the mop.

However, many of the species considered more difficult, particularly *Diapterons* and some *Mesoaphyosemion*, seem to reproduce more comfortably using the **Sphagnum Moss Method.** In this case, it is important to monitor the water quality because the sphagnum will change the pH, sometimes dramatically.

Most members of this group learn to take flake foods, but for optimum breeding condition, live foods are suggested, particularly white or black worms. *Diapterons* seem particularly fond of *Daphnia*. Frozen foods are taken readily.

Aphyosemion striatum Lamberene
Photo: Mike Jacobs

Aphyosemion ogoense pyrophore
Photo: Richard Pierce

Aphyosemion elberti Ntui
Photo: Mike Jacobs

Aphyosemion bitaeniatum ljebu-Ode
Photo: Richard Pierce

KN PRESS - T.R. GRADY

Epiplatys: The Hunters

Surface dwelling predatory killifish is a good starting point to consider for this group, however it is not completely true. Members of this group of fish inhabit rivers and streams primarily in all biotopes from savannah to rainforest and even along the coastal plains. They range from the headwaters of the Nile River to Gambia and southward into the Congo.

One of the most desired and gentle of the killies is found here, the clown killie - *Epiplatys annulatus*. Looking at most members of this group can remind an aquarists of small pike, a nasty predator of the North American rivers and lakes, but there is no relationship beyond shape.

For the most part, *Epiplatys* should not be considered as good community fish to mix in a general tropical fish population.

Coloration of *Epiplatys* is different than the previous groups. Black vertical bands are common in most species and nearly every female of the species have these bands. Colors range from quite beautiful oranges and yellows, to somewhat duller olive and tan. Some species have a number of scales that are edged in a reflective cooler, generally gold or green.

Most members of the *Epiplatys* group are moderately easy to induce to spawn and are considered to be floating mop spawners. Generally the eggs are found high in the mops. Eggs tend to hatch in two to three weeks and most fry (excluding *E. annulatus*) easily take brine shrimp nauplii immediately, but infusoria is never a bad idea to add to the early diet.

In nature, *Epiplatys* are know for eating flying insects and and do leap from the water to catch prey. In the home aquaria, *Epiplatys* are not finicky eaters and will take most foods offered. However, in particular, *E. annulatus* needs small food, baby brine shrimp and infusoria are preferred. Fry may not be large enough to eat baby brine shrimp, but a heavily planted tank provides enough microorganisms.

Epiplatys togoliensis

Photo: Tom Grady

Epiplatys fasciolatus

Photo: Tom Grady

Epiplatys infrafasciatus Zenkeri Route *Dehane ADK 09-297*

Photo: Mike Jacobs

Fundulopanchax: Switch Things Up

Fundulopanchax a large group of killifish that seem to fit somewhere between annual and non-annual species and do take on some characteristics of both.

This group of killifish reside primarily in caostal regions of West Africa from Togo to Cameroon and one group, *Raddaella*, are gound on the plateau of Gabon and Cameroon.

Currently the larger group *Fundulopanchax* is split into five sub-groups: *Fundulopanchax* Myers, 1924, *Gularopanchax*, Radda 1877, *Paludopanchax* Radda 1977, *Paraphyosemion* Radda 1977 and *Pauciradius* Wildekamp and Zee, 2005. Fundulopanchax are discussed in greater depth in **a** future volume of **The Encyclopedia**.

One of the biggest problems within the hobby is the misidentification of many members of the *Paraphyosemion* group, more commonly known as the "**Gardneri Group**". Because there have been so many different populations of these species brought into the hobby, many populations have either been lost, or hybridized. It is essential for hobbyists to understand the separation and differences in each of the species and make certain they are properly identified.

Overall, *Fundulopanchax* species of the *Gularopanchax* tend to be physically larger, with some species reaching six to seven inches. Species such as *Fp. filamentosa* and *Fp. robertsoni* are smaller, but the extensions on their fins could easily be considered plumage.

Most will spawn in breeding mops and there is a tendency to find eggs in the middle to lower areas of the mops. However, the use of peat moss in bottom containers works very well and in some cases is the preferred technique for certain species. Most eggs can easily be water incubated, but some species may take six to eight weeks in water. Some hobbyists prefer to place those eggs on wet pet moss for a similar time period.

Fundulopanchax arnoldi Ughelli
Photo: Richard Pierce

Fundulopanchax amieti
Photo: Tom Grady

Fundulopanchax gardneri Aquarium strain
Photo: Mike Jacobs

(above) *Fundulopanchax sjoestedti* is best known as the Blue Gularis and is also the emblem fish for the American Killifish Association. (below) *Fundulopanchax walkeri* GH2 Kutense (top next page) *Fundulopanchax deltaense* **Photos: Mike Jacobs**

Roloffia: Controversy Personified

The *"Roloffia"* hold a unique position in the aquarium hobby as well as the scientific community. There is a real division over the proper name identification that should be used and hobbyists as well as taxonomists are divided on this subject. Officially, the name *Roloffia* does not exist according to the ICZN, On a genus level everyone is in agreement, but it is the 'common name' to identify the group . Unfortunately this often confuses hobbyists when communicating from country to country worldwide.

In tems of habitat, *"Roloffia"* replace *Aphyosemion* north of the Dahomey Gap, a geographical savanna corridor that interrupts the West African rain forest in Benin, Togo and Guinea. North of this dry zone is known as the Upper Guinean Forest and to the south is the Congolian Forest Zone. At one time the Dahomey Gap was epicontinental sea (Collier & Murphy[1]) that forced separate lines of evolution.

Current literature designates 'Roloffia' as closer to **Epiplatys** than to *Aphyosemion* and are contained in the Subfamily *Epiplatinae* and hold a position known as *Callopanchini*. The three genus are *Callopanchax*, *Archiaphyosemion* and *Scriptaphyosemion*. *Nimbapanchax* is a subgenus of *Archiaphyosemion*.

Most members of *"Roloffia"* are stunning with green to blue bodies covered in red spots. The fins contain large red spots and bands. Most exhibit yellow, blue or white marginal bands and red submarginal ones in the anal and caudal fins.

Overall the major distinction between the groups is size and breeding techniques. *Callopanchax* are among the larger killies and most hobbyists prefer to use annual techniques to breed the pairs and incubate the eggs which can take up to six months. The smaller members of *Archiaphyosemion* and *Scriptaphyosemion* are bred in floating mops, much in the same manner as Aphyosemion, and the eggs are incubated in water for about three weeks.

Feeding these species varies a bit. The smaller species are fine with pretty much anything. *Callopanchax* will relish small red worms. Standard frozen foods and live foods will be taken and flake foods can be offered, but may require some starvation for the fish accept.

KN PRESS - T.R. GRADY

Nothobranchius: African Survivors

Some of the most beautiful freshwater tropical fish in the world just happen to be killifish and in particular the group identified as *Nothobranchius*. It has been said *Nothobranchius rachovii* is the most attractive fish of all, but there have been other species recently discovered that certainly can challenge that claim.

A look at a map detailing the savannas of Africa will pretty much trace the range of the habitats of *Nothobranchius* and its related species. Although none are found on the island of Madagascar.

Although Nothos contain several subgroups, those will be discussed in much greater detail in a separate volume of The Encyclopedia of Killifish. Species of *Aphyobranchius*, **Fundulosoma**, *Pronothobranchius* and *Paranothobranchius* are all subgenus names within this section and generally identified in the hobby as *Nothobranchius*.

Nothos are all annual species of killifish, which means the environmental factors of wet and dry seasons determine the length of their viable lifespan. Over thousands of years, *Nothobranchius* have adapted to these shortened cycles with rapid growth and a strong urge to reproduce as soon as possible. Some species begin sexing out in six to eight weeks after hatching and are fully capable of reproducing at that point. Some areas Nothos inhabit may only have water for less than six months and dry periods that last more than a year. Yet they have adapted survival mechanisms to survive those extremes.

In the home environment, hobbyists have had to develop methods to simulate the natural cycles. For most species this simply means they are allowed to spawn in a bottom substrate which is then dried and packed away for extended periods of time - anywhere from 2 months to two years in some very rare cases. This is discussed in the section on breeding killifish and that will be expanded in **The Encyclopedia**

The green surrounding the dark center of the Congo Basin is where most species of Nothobranchius are found.

(above) *Nothobranchius taeniopygus* Itigi
(below) *Nothobranchius rachovii* Beira
Photos: Mike Jacobs

volume detailing Nothos.

One important consideration for maintaining Nothos at home is the use of up to a tablespoon of aquarium or sea salt in the tanks. Nothos are susceptible to velvet and some other diseases and organisms that salt will inhibit. Fry are exceptionally susceptible to Velvet

In addition to Velvet, the internal parasite microsporidium known as *Glugea anomala* is a serious condition that seems to affect Nothos more than any other group of fish except perhaps sticklebacks. Initially recognized in the early 1990s, this parasite was known as the Notho Disease and considered incurable. Today it is controllable using certain medications including Flubendizole™, but infected fish should be destroyed and any eggs need to be treated (* see the section on diseases) upon hatching to prevent the potential spread of this condition.

This group of killies currently contains over 70 described species, numerous unidentified species and a number of questionable populations that may or may not be individual species. Several new species are identified each year.

For the novice Notho-keeper, perhaps the first species should be *N. guentheri* or *N. foerschi*. These two species are relatively easy to breed and raise and both are quite beautiful with deep red tails and bright blue bodies. The idea is for the novice hobbyist to have initial success before moving on to more sensitive or difficult species. Since both of these Nothos have incubation periods of two to three months, the aquarist does not have to wait long periods of time before experiencing the results of their efforts.

Nothos do prefer live foods and, while they can be coaxed into taking dry foods with starvation, they may ignore the flakes completely. For the most part, white and black worms are relished. Mosquito larvae will be consumed like nothing else. All species will accept frozen alternatives easily.

Nothobranchius makondorum Nakapanya
Photo: Tom Grady

Nothobranchius rubripinnis
Photo: Tom Grady

Nothobranchius foerschi
Photo: Tom Grady

KN PRESS - T.R. GRADY

Pachypanchax: The Isolated Killifish

A small group of distinct killifish inhabit the island just off the east coast of africa, Madagascar and a small island to the north - the Seychelles. All members of the *Pachypanchax* are considered threatened in the wild.

For the most part, members of this group are found in the rivers and streams that come from the higher elevations on the island, but a couple of species are found in the brackish waters where the rivers empty into the ocean.

With the possible exception of *Pachy. playfairi*, most species of *Pachypanchax* grow to 3 to 4 inches.

Breeding *Pachypanchax* is not difficult. They are plant spawners and will lay many eggs in floating spawning mops. Eggs generally take 2-3 weeks to hatch and the fry easily eat brine shrimp naupalii.

Pachypanchax arnoulti
Photo: Richard Pierce

Pachypanchax omolonotus
Photo: Tom Grady

The island of Madagascar is home to many isolated species of animals and fish which evolved separate from those on the Adrican continent. Lenurs are a good example, but killifish and cichlids also demonstrate the same isolated evolution.

Pachypanchax playfairi
Photo: Tom Grady

Lampeyes: The Lanterns of Africa

Lampeyes are the schooling members of the killifish and vary in size from some of the smallest fish (*Aplocheilichthys normanni*) to some of the largest species (*Lamprichthys tanganicanus*). The one distinction of all Lampeyes is the light blue reflective scales above their eyes that gives the appearance of tiny glowing fireflies in a fish tank.

For the most part, Lampeyes inhabit the upper regions of the water column and are in constant motion. In large environments they remain close together to form a school.

Lampeyes are found in nearly every country in Africa and inhabit some of the same waters as *Nothobranchius* and *Aphyosemions*.

Most species are not difficult to induce to spawn and use of floating mops in the upper reaches of the tanks is the primary method. Because the eggs and fry of many species are so small, some hobbyists remove the mops from the breeding tanks and place those in small tanks until all the eggs hatch. One of the benefits of this is infusoria is already in the mop strands and for the tiny fry the microorganisms assist in raising them to a size where baby brine shrimp can be taken. If the hobbyist has a *Paramecia* or *Euglena* culture, these are very useful for the tiny young.

Lampeyes can be easily taught to take flake foods, but they relish the smaller live foods. Even brine shrimp nauplii are accepted as a base food for many of these species. Grandal worms and microworms are also good options.

There are a number of genera and sub-genera and some are quite different in coloration or body shape. Lampeyes are actually members of the Family *Poeciliidae*. Included in this group are members of the *Aplocheilichthys*, *Procatopus*, *Micropanchax* and *Fluviphylax* groups. Each of these groups have been further split into several subgenus.

Aplocheilichthys luxophthalmus
Photo: Richard Pierce

Plataphocheilus cabindae GIS 00-21
Photo: Richard Pierce

Procatopus abberans Muyuka
Photo: Gary Elson

- 91 -

KN PRESS - T.R. GRADY

Killifish of South America

If there is wide diversity in the killifish of Africa, those residing in South America are no less fascinating and the variety of species is amazing. There is no country on this continent that does not offer something unique in a diverse distribution that ranges from dry, arid regions around Lake Maracaibo in Venezuela to the depths of the Brazilian Amazon Rainforest. Killies can be found in Lake Titicaca, the highest navigable lake in the world nearly 2-1/2 miles above sea level in the Andes. Sadly, many species of Lake Titicaca's *Orestes* have gone extinct in that lake due to introduction of trout and silversides.

In a sense South American Killifish mirror their African cousins in many ways. There are numerous annual species found throughout the continent that require varying periods of dry incubation. These are even more diverse than the *Nothobranchius* of East Africa and the several different genus and subgenus groupings contain both very large and tiny species. On the other hand, the *Aphyosemion* and *Fundulopanchax* of West Africa are replaced by a wide variety of *Rivulus* in all forests and rainforests throughout South America and nearly every country that touches the Amazon basin has its own groupings. Even the mangroves along the northern coasts and parts of Brazil contain some fascinating hermaphroditic *Rivulus*. *Rivulus* provide the vast majority of killies found on the continent and exist in every river basin.

It is almost easy to see where the divisions of killies occurs in South America by looking at the climate map. With the exception of *Rivulus* which are present nearly everywhere, the groupings of fish tend to follow the climatic regions, with specific annual species found in the regions with distinct wet and dry seasons and everything else in the tropical and temperate areas.

While there is some crossover of annual

species in each of the seasonal dry regions, a pattern can be seen with *Austrolebias* species focused in the Temperate and Cold areas and *Simpsonichthys* more often found in river basins in the Monsoon climate.

In the northern part of South America, annuals with extended incubation periods are found in the Steppes and Monsoon areas as well as around Lake Maricaibo in Venezuala.

Gnatholebias hoignei   **Photo: Tom Grady**

South American Annuals

As stated in the overview of South American Killifish, there is a huge diversity of annual species of killies that can be found in different sections of South America. Some of these species are unique in both coloration and physical shape. One of the most obvious differences between African and South American annuals is that most S.A. annuals have points and extensions on the dorsal, anal and caudal fins whereas their counterparts tend to be far more rounded. But that is only one of many differences. South American annuals present very different color patterns, including bars and stripes nearly nonexistent in most species of Nothos. Despite the differences, the beauty of many *Simpsonichthys* rivals that of the *Nothobranchius*.

One absolute similarity between the annuals of the two continents is behavior. The fish still lay their eggs in places that dry out and must go through an incubation period to hatch (although some experiments indicate certain species of annuals can be water incubated to a successful hatching).

It is of particular interest to note that evolution has created different groupings based on what river basin or savanna region the animals originated. The region surrounding Lake Maracaibo in Venezuela is home to larger, deep-bodied species of *Austrofundulus* and *Gnatholebias*. The regions south and east of the Amazon basin from Brasilia to Rio De Janeiro and Sao Paulo north to the Recife region contain dozens of species of *Simpsonichthys* and *Cynolebias*. Perhaps the most fertile region for annuals in South America is the Sao Francisco river system with new species being discovered on a regular basis.

Further south *Simpsonichthys* are overtaken by species of *Austrolebias* which join *Cynolebias* and *Megalebias* in southern Brazil and continue though Uruguay and into Argentina. To the

Simpsonichthys carlettoi

Photo: Richard Pierce

KN PRESS - T.R. GRADY

west, including Bolivia and Paraguay and even north into Peru can be found other groups of larger annuals, *Pterolebias*, *Neofundulus* and *Trigonectes* to mention a few. There are many sub-groups within all of the above mentioned families.

To breed nearly all of the South American annuals is not much different than the techniques used for Nothos. Some sort of size appropriate container containing peat moss or coconut coir can be used and examined for eggs every week or two. It should be noted some species of South American annuals prefer to dive deep into the breeding media and the use of deeper containers is indicated. This will be covered in greater depth in the upcoming volume on South American Annuals.

The peat moss or coconut coir is processed and dried for storage in the incubator for periods of three months up to one year or more depending on the species.

South American annuals are not generally selective eaters and take almost any live or frozen food and can also be trained to accept flake foods.

Neofundulus splendidum
Photo: Richard Pierce

Pituna compacta

Photo: Richard Pierce

Pterolebias phasianus
Photo: Richard Pierce

Rivulus: The Ubiquitous Residents

Rivulus are perhaps the most widespread group of killies and range from North America, through the Caribbean basin, the length of Central America and nearly anywhere there is water in South America. Contained within this group is the hermaphroditic species *Riv. marmoratus* as well as the incredibly beautiful *Riv. Xiphidius* and the large *Riv. hartii*. *Rivulus* are found in saltwater and fresh. They are in rivers, streams, brooks and ponds. Mangroves are home to several species and others inhabit specific islands in the Caribbean and found nowhere else.

Essentially *Rivulus* inhabit the similar ecological environments as the *Aphyosemion* and *Fundulopanchax* in Africa.

They are the middle to top water species that use floating plants as hiding and breeding places. Many species have a strong jumping tendency and tanks must be well covered or toastier snacks will be found on the floor. In nature the jumping behavior is an escape mechanism to avoid predators such as cichlids and numerous other large, snack hungry species.

Not long ago, every *Rivulus* was identified as just one genus, but over the past several years, taxonomists have split the family into a number of subgenus. Despite the scientific changes, most hobbyists still tend to refer to all of the fish in this group as *Rivulus*. This is beginning to change and aquarists should become aware of the differences.

Breeding: For the most part, they spawn fairly readily in floating mops. Many species will attempt to place the eggs right at the water surface and in a few cases above. Eggs are often found in the tightest strands of yarn near where it is tied. Eggs are even found attached to the flotation anchor. Normally it takes two to three weeks for the eggs to hatch and nearly every species has no problem eating brine shrimp nauplii immediately. It never hurts to add infusoria to the fry trays (which should

Rivulus xiphidius

Photo: Richard Pierce

Rivulus agilae Akoloi

Photo: Richard Pierce

Rivulus madiaensis

Photo: Mike Jacobs

- 95 -

have plants or algae already).

Rivulus do take some time to mature. It is rare to find any species that sexes out at anything less than six months and many species are longer. *Rivulus* can live for a long time in the home environment, sometimes six or more years. They are also quite adaptable to poor conditions, weathering everything from pollution to periodic dry periods in nature. Some species of *Rivulus* have been known to travel across dry land to seek out new pools of water.

Orestias - Almost Unknown Killifish

Members of the *Orestias* family are probably the least known killifish. They are mentioned here just to acknowledge the group.

Most originate high in the Andes Mountains of Peru, Ecuador, Bolivia and Chili and are extremely difficult to transport because of fragility to pressure as well as a need to be maintained in cool water. There are no species currently known in the hobby although a few have been captured for study and reside primarily in Europe. As a group, there are 45 species.

Although this group originally stood on its own, today they are considered Pupfish which also contains *Cyprinodon* in North America and the *Aphanius* of Europe and the Middle East.

Most, if not all, members of *Orestes* are considered threatened with extinction. The introduction of a variety of fish for a combination of sport fishing and food consumption has decimated natural populations in Lake Titicaca. There used to be a number of species in Lake Titicaca of Peru, but most are now considered extinct. Orestias cuvieri has not been seen since 1939. The river basins surrounding the lake may contain some species.

However, the genus extends throughout the Andean Altiplano from northern Peru south to Antofagasta in Chili with 19 species identified in ponds and lakes not within the Lake Titicaca region. There are a total of 42 recognized species of which 23 are in rivers and streams around Lake Titicaca.

There is very little information on reproduction of *Orestias* species, but it can be assumed they spawn similarly to most other pupfish. Because there is no information on their feeding habits, it also has to be assumed they will feed on most live foods small enough to be eaten.

The Other South Americans

There are a few other species found in South America that should be noted if only for their unique status.

Fluviphylax are considered the 'Lampeyes' of the Amazon, holding a similar niche to those found in Africa. There are five known members of this small group of miniature killies which contains the smallest killie currently known, *Flu. palikur,* which reaches the maximum size of less than 1/2 inch. All species are tiny and are endemic to the Amazon Basin.

Fluviphylax are all considered surface dwelling species. Based on their small size, it can be assumed their primary source of nourishment is infusoria and tiny larvae and crustaceans. Little is known about their breeding.

Their aquarium status is unknown. It is possible a few pairs exist, but if so, it is likely they are only for scientific study.

The members of this group include: *Flu. palikur, Flu. obscurus, Flu. pygmaeus, Flu. simplex* and *Flu. zonatus.*

Orestias cuvieri:

North\Central America and the Caribbean

North America covers everything from Panama to the Arctic Ocean above Canada, the United States and all of the islands in the Caribbean. The diversity of killifish is fascinating and there is a real crossover with many species of *Rivulus* ranging north from South America into Central America and across many islands including Cuba, Trinidad and Martinique. Many of the species are unique to their locations.

With the exception of one interesting fish, *Millerichthys robustus* that is found in Mexico, there are no annuals in this region.

Central America from Panama through Costa Rica is home to a latge number of diverse Rivulus that are still being discovered. Recent collections in Panama have added a number of new species to science and the hobby.

Cyprinodon alvarezi is extinct in the wild since the 1990s, but captive populations exist. Originally it came from singlo location, El Potosí in Nuevo León state, northern Mexico.

Photo: Richard Plerce

United States, Canada and Mexico

North America is also home to some of the most endangered killifish in the world - the Pupfish of the American Southwest and throughout Mexico. Most of these species are protected within their natural ranges and nearly all require a permit to keep in the home aquaria. Despite the restrictions, there are

a few captive populations maintained in the hobby, even some of species now considered extinct in the wild.

The diversity of species and the conditions they reside in North America is interesting. Killies range from the cold and frozen northern parts of the United States and into Canada to

KN PRESS - T.R. GRADY

the hot rainforests of Central America. Some species are found in the salt and brackish water of nearly all coastal areas and islands. Then there are the Pupfish of the nearly arid regions of the United States and Mexico which survive in isolated springs.

Most hobbyists in the U.S.A. can step out their back door to a small pond or stream and find one type or another killifish. While these are primarily a species of *Fundulus*, in some regions, a real mix is available including a pupfish, *Cyprinodon variegatus*, and a small delicate beauty, *Leptolucania parva*. *Fundulus diaphanus is* ubiquitous to most of Eastern U.S. from North Carolina to Canada and across the Great Lakes. The mummichug, *F. heteroclitus*, can be found in nearly every brackish waterway from Maine to Florida and is the fish that most likely originated the name 'killi'fish.

Florida is home to a large number of killifish and a few are solitary species or genus. The American Flag fish, *Jordanella floridae* is everywhere and the variation in patterns and colors is interesting. A number of *Fundulus* can be found across that state and there is a saltwater species, *Floridichthys carpio,* that is rarely seen in the hobby.

Actually, the entire Southeastern USA and well into the Midwest, there are many species of killifish including some monster sized fish like *F. catenatus* and the stunning *F. zebrinus*.

Because of the diversity of habitats, there is no one way to breed most North American species. For most *Cyprinodon* species heat plays a important role with temperatures rising well into the 90s Fahrenheit (30º+C).

Many of the species found in the southern regions of the USA are typical plant spawners and eggs can be treated like other mop spawners. Yet the American Flag Fish, *Jordanella floridae*, displays nesting behavior.

Feeding North American killies is rarely a problem. Most will take any type of frozen food and certainly live foods are on the menu. With

*Rivulus weberi - **A species from Panama**
Photo: Mike Jacobs*

time they learn to take flake foods.

Cyprinodon: The Pupfish

The familiar name of Pupfish is an overall identifier for a number of genus of killifish that range around the world from the American Southwest to South America and into the Middle East, but most often are focused on members of the *Cyprinodons* of Mexico and the United States. These are generally small, deep-bodied fish that are isolated in range and found in some of the most extreme conditions. The heat of Death Valley can exceed 100ºF (38ºC) where a number of endangered species are found. Alternatively Pupfish are found throughout the Caribbean basin - in most Central American countries as well as on a number of islands. A variety of forms of *Cyprinodon variegatus* are comfortable in the salt and brackish waters of tidal pools and river estuaries. Several species have evolved to fill interesting niches in the different locations.

Pupfish make for fascinating viewing. They are very active, displaying for and chasing each other constantly. The vast majority either have blue scales or blue reflective edges on the flanks and a variety of black markings in the fins. In nature, many species live in springs and

Ash Meadows in Nevada is a U.S. Fish & Wildlife Refuge for several species of Pupfish. Above is a picture of the area demonstrating the harch climate that surrounds the springs. Below is the actual spring where *Cyp. nevadensis* survives in small numbers. To the left is a male *Cyp. nevadaensis* hovering over the algae beds it feeds on. Photos: Tom Grady

may be the only fish in an isolated location. Their biggest enemies are birds.

Now for the bad news. Since nearly every species, with the exception of *C. variegatus* is considered at least threatened, if not endangered, they are generally illegal to maintain as aquarium fish. Stories and theories abound about 'captive' populations, but the trade, sale or display of those species is prohibited. There are a couple of exceptions and, strangely enough, those species are extinct in nature. Since the fish no longer exists in the wild, it can be kept. The other exception is a qualified aquarist can arrange for a permit through the U.S. Fish and Wildlife Department. Those are nearly impossible to obtain.

As uncommon as *Cyprinodon* species are in the home aquarium, they are not difficult to breed if given the proper environment. Depending on the particular species, temperature plays a large part in the process, with some members of this family requiring a temperature of 90ºF+ to initiate spawning behavior. Most lay eggs in bottom mops and the eggs are no different than those of other killifish, hard shelled and can be handled physically. Incubation varies with species, but two to three weeks is the norm.

Most members of *Cyprinodon* feed on crustaceans - *daphnia*, and *cyclops* - in the wild, but they also graze on algae and it is an important part of their nutrition. Frozen foods will work well as do most cultured live foods.

***Cyprinodon nevadensis* shoshone is one of the endangered species found at the Ash Meadows National Wildlfie Refuge in Nevada. This refuge also works with the Devils Hole Pupfish, *Cyprinodon diabolis*.**

Photo © Anthony Terceira

Fundulus Are Everywhere

Fundulus, often called topminnows locally, are found throughout the North American continent with the vast majority of species located in the United States. A few species are located in the Caribbean Basin in isolated pockets. There is no single habitat which can be specifically designated for members of this group. Northern species can be found in large lakes that freeze in the winter and others are in ditches in semi-tropical regions. *Fundulus grandis* inhabits the salt waters of the Gulf of Mexico from Texas to Florida and even in Cuba. It is replaced by *Fundulus heteroclitus* along the Atlantic coast as far north as Maine. Yet *Fundulus diaphanus* is a completely freshwater species found throughout most of the eastern parts of the U.S. and into the Midwest. There are populations of this species in Canada. *Fundulus zebrinus* is a striking species of this group that can be found from Montana to Texas.

The *Fundulus* also contains one of the monster killies, *Fundulus catenatus,* the Northern Studfish, which grows to six or more inches., while most other species rarely exceed about four inches. *Fundulus chrysotus*, found primarily in ditches in Florida is intriguing, presenting with both normal and melanistic (black spotted) forms.

Fundulus tend to fill the same niche as the *Rivulus* of South America and the *Aphyosemion* of Africa.

All are plant spawners and can be offered floating mops that stretch from top to the bottom of the breeding tanks. Eggs hatch in about two weeks.

Feeding these species demonstrates no special requirements and most *Fundulus* will take most offerings from live foods to flakes. In the wild they are insectivores, but also feed on crustaceans, worms and insect larvae. It's not unknown for them to eat fry.

Fundulus chrysotus

Photo: Mike Jacobs

KN PRESS - T.R. GRADY

The Caribbean Basin Killies

There are a number of killifish covering many families found in both the South and North American continents including numerous species of *Rivulus* and *Cyprinodon*. There are a few isolated genus found in the Caribbean including two species of pupfish, *Cubanichthys*, one in Cuba (*Cub. cubensis*), the other in Jamaica (*Cub. pengelleyi*). The *Cyprinodon* species identified as *Cyp. sp.* Caymon is found on Grand Caymon island in an area known as Hel. It is likely a variation of *Cyp. variegatus*.

However, *Rivulus* are spread throughout the Caribbean Basin and heavily populate Central America from Panama to Mexico.

Rivulus marmoratus, in spite of it's lack of color except brown, may be found in nearly every mangrove and *Cyprinodon variegatus* is well known as a coastal species. There are two species of *Rivulus* in Cuba *R. cylindraceus* and the newly described *R. berovidesi* and another on the island of Martinique, *R. cryptocallus*. To the north of South America, Trinidad is home to *R. hartii*.

Coming from the north, many species of *Cyprinodon* are found throughout Mexico and on at least a few islands in the Caribbean. In fact the Bahamian island of San Salvador is home to three unique species of *Cyprinodons*.

Bermuda is home to a *Fundulus* species, *F. bermudae* and one other possible species, *F. relictus*. However recent studies may separate some populations of these species into two or more. F. grandis is found on several islands. Pupfish feed in the muck on the bottom of cost habitats and seek out a variety of foods ranging from certain types of algae to fish eggs and crustaceans. However the *Rivulus* are focused more on insects and larvae and spend more time near the surface. In the home aquaria, most accept the common frozen and live foods and some will take flake foods.

Breeding varies with the species, but either floating mops or bottom mops will work.

Rivulus cylindraceus **Al Castro** is found on the island of Cuba. There have been a couple color variations over the years.
Photo: Mike Jacobs

Rivulus uroflammeus siegfredi comes from Panama
Photo: Tom Grady

Rivulus caudofasciatus is found in the northern regions of South America.
Photo: Tom Grady

Other North American Species

There are a number of killies in North America that cannot be identified with one group or another. These include *Lucania, Adinia, Leptolucania*, and several others. Each contains only one or two species, but in many cases are popular with hobbyists.

Lucania goodei (Bluefin Killie) and *Lucania parva* (Rainwater Killie) are the only two species in this group. *L. goodei* is a fairly popular killie and is identified by the light blue coloration of its anal and dorsal fins. There are a variety of populations which exhibit some red in the caudal and anal fins, while others show practically none. L. parva exhibits little color other than an olive to silver body. Occasionally there is a little red in the anal fin. L. goodei is occasionally found as a stowaway in bags of other fisg sent from Florida fish farms to wholesale distributors or aquarium stores.

Both prefer small crustaceans as a food, but will take most offerings of frozen or live foods.

Leptolucania ommata stands by itself and is a swamp killie found in Florida and Georgia. This is one of the two smallest fish found in United States rarely exceeding one inch in size. Generally *L. ommata* is a golden yellow fish with two darker spots in the cataleptic. There may be some greenish reflectivity to the body. This species is highly adaptable and makes for a beautiful part of a "swamp" environment along with *Herterandria formosa*, some pygmy sunfish species and swamp darters. Basically they eat any small live food and prefer to live in heavily planted tanks with floating plants. They spawn in the plants or a floating mop.

Adinia xenica (Diamond Killie) is a small killie that rarely exceeds two inches that is found along the west coast of Florida and across the gulf coast to Texas. They are a brackish water species that can be adapted to freshwater. They are mid-water breeders and a floating spawning mop is fine. Like all killies, feeding them consists of live and frozen foods.

Carmanella pulchra is found on the Yucatan Peninsula and is generally a small fish, less than 2 inches. Whether or not it has been kept in the hobby is questionable. There is little information on breeding it. It does feed on the traditional foods of small crustaceans, *Daphnia* and *cyclops*, according to stomach contents.

***Lucania goodei* Corporate Park Road**
Photo: Richard Pierce

***Jordanella floridae* -** The American Flag Fish
Photo: Mike Jacobs

Adinia xenica   Photo © Anthony Terceira

Euro-Asiatic & Middle East

There are only two groups of killifish that span across the Euro-Asiatic region. Species related to *Aphanius* Nardo 1827 cover a few genera that range from Spain through North Africa and into Turkey and Iran.

Closely allied with the *Cyprinodon* of North America and to a lesser degree the *Orestias* of South America, the name *Aphanius* had undergone some controversy in recent years that was finally settled by the ICZN in 2003 and the name *Lebias* was invalidated.

The other group, Aplocheilus, primarily inhabit the sub-continent of India and a few nearby areas. These are pike-like fish with elongated bodies and mouths that look like they are ready to attack at any moment. For the most part, *Aplocheilus* inhabit the upper regions of the waterways.

Aphanius & Valencia

Aphanius can be found on both sides of the Mediterranean Sea from Spain to North Africa as well as into Turkey and Iran. For the most part, these killies are very tolerant of the salinity of water and can be found in everything from fresh to brackish to salt waters. Most often they reside in slow moving streams and ditches near coastal waters.

There are over forty species currently identified and new ones are being described on a regular basis. A number of species are seriously threatened due to limited distribution. Also within the same group are several subgenus. *Anatolichthys* Kosswig & Sozer 1945, **Turkichthys** Ermin 1946, Kosswigichthys Sozer 1942 and Tellia Gervais 1863

Similar to *Aphanius*, if only by shape and geography are three species of *Valencia*, *Val. hispanica*, *Val. letourneuxi* and *Val. Robertae*, all found along the Mediterranean coast of Spain to Greece and Albania. These species are considered highly endangered and are protected.

General care for members of these groups requires a small amount of salt added to the water.

All are fairly easily fed with live and frozen foods and over time can be induced into taking flake foods.

Aphanius sirhani   **Photo: Richard Pierce**

(above) *Aphanius fasciatus* maschio di Valano   (below) *Aphanius sophiae*
Photos: Stefano Valdesalici

Aplocheilus

The sub-continent of India and the island of Sri Lanka gives the hobby a single group of killifish, the *Aplocheilus*. Contained in this group are eight species of easily maintained surface dwelling killies. For the most part, members of *Aplocheilus* are found in most slow moving waters throughout its range.

Aplo. lineatus is often found in pet stores under the name Golden Wonder Killie. Occasionally *Aplo. panchax* (The Blue Killie) makes an appearance. The other members are extremely uncommon in the hobby. This is not so much due to difficulty in breeding and maintenance as much as it is the limited interest in some species. Other than *Aplo. lineatus* and *Aplo. kirchmayeri*, most species are not particularly colorful. However, all have been in the tanks of killie-keepers at one time or another.

In the wild, members of this group are primarily carnivores and feed on insects and their larvae as well as worms. In the home aquarium, *Aplocheilus* are not finicky eaters and will take any frozen or live food greedily. They even attack flake food.

All of the *Aplocheilus* are fairly easy to maintain in the home aquaria. These are surface killies, often found in the upper reaches of the aquarium hiding near floating plants.

All spawn in floating mops and will produce quite a few eggs in a 24-hour period. Eggs take roughly 18 days to hatch. The fry are large enough to take newly hatched brine shrimp.

Are Rice Fish Killies?

Over the years, several groups of fish have been said to be closely related to killifish. Among them are Oryzias - the Rice Fish of the Far East and Japan. There are many similarities, both in the niche they hold in nature as well as their reproductive habits. The same could be said for Raindow fishes and Blue Eyes, but technically they are not killifish.

However, many hobbyists maintain these species because of the close relationships.

Aplocheilus lineatus Bold
Photo: Richard Pierce

Aplocheilus panchax
Photo: Mike Jacobs

Aplocheilus kirschmeyeri
Photo © Anthony Terceira

Appendix A - Taxonomic Listing of Killifish

Order *Cyprinodontiformes* Berg 1940
    Suborder *Aplocheiloidei* Bleeker, 1859
        Family *Aplocheilidae* Bleeker, 1859
            Genus *Aplocheilus*
            Genus *Pachypanchax*
        Family *Nothobranchidae* Garman 1895
            Genus *Nothobranchius*
                Subgenus *Nothobranchius*
                Subgenus *Adiniops*
                Subgenus *Aphyobranchius*
                Subgenus *Paranothobranchius*
                Subgenus *Zoonothobranchius*
            Genus *Fundulosoma*
            Genus *Pronothobranchius*
            Genus *Aphyosemion* Myers 1924
                Subgenus *Aphyosemion* Myers 1924
                Subgenus *Chromaphyosemion* Radda 1971
                Subgenus *Diapteron* Huber & Seegers 1977
                Subgenus *Episemion* Radda & Purzl 1987
                Subgenus *Iconisemion* Huber 2013
                Subgenus *Kathetys* Huber 1977
                Subgenus *Mesoaphyosemion*
                Subgenus *Raddaella* Huber 1977
                Subgenus *Scheelsemion* Huber 2013
            Genus *Foerschichthys* Scheel & Romand 1981
            Genus *Fundulopanchax* Myers 1924
                Subgenus *Fundulopanchax* Myers 1924
                Subgenus *Gularopanchax* Raddda 1977
                Subgenus *Paludopanchax* Radda 1977
                Subgenus *Paraphyosemion* Radda 1977
                Subgenus *Pauciradius* Wildekamp & Zee 2005
            Genus *Fenerbahce* Ozdikmen, Polat, Yylmaz & Yazycyodlu 2006
            Genus *Epiplatys* Gill 1862
                Subgenus *Epiplatys* Gill 1862
                Subgenus *Lycocyprinus* Peters 1868
                Subgenus *Parepiplatys* Clausen 1967
            Genus *Aphyoplatys* Clausen 1967
            Genus *Pseudepiplatys* Clausen 1967
            Genus *Callopanchax* Myers 1933
            Genus *Archiaphyosemion* Radda 1977
                Subgenus *Archiaphyosemion* Radda 1977
                Subgenus *Mimbapanchax* Sonnenberg & Busch 2009
            Genus *Scriptaphyosemion* Radda & Purzl 1987

        Family *Rivulidae*
            Genus *Rivulus*
                Subgenus *Rivulus* .s.s Poey 1860
                Subgenus *Anablepsoides* Huber 1992

Subgenus *Atlantirivulus* Costa 2008
Subgenus *Benirivulus* Costa 2006
Subgenus *Cynodonichthys* Meek 1904
Subgenus *Laimosemion* Huber 1999
Subgenus *Melanorivulus* Costa 2006
Subgenus *Oditichthys* Huber 1999
Subgenus *Owiyeye* Costa 2006
Subgenus *Volerivulus* Fowler 1944
Genus *Prorivulus* Costa 2004
Genus *Aphyolebias* Costa 1998
Genus *Austrofundulus* Myers 1932
Genus *Austrolebias* Costa 1998
Genus *Campellolebias* Vaz-Ferreira & Sierra 1974
Genus *Cynolebias* Steindachner 1877
Genus *Cynopoecilus* Regan 1912
Genus *Kryptolebias* Costa 2004
Genus: *Leptolebias* Myers 1952
Genus: *Notholebias* Costa 2008
Genus *Maratecoara* Costa 1995
Genus *Megalebias* Costa 1998
Genus *Micromoema* Costa 1998
Genus *Millerichthys* Miller 1892
Genus *Moema* Costa 1989
Genus *Nematolebias* Costa 1998
Genus *Neofundulus* Myers 1927
Genus *Papiliolebias* Costa 1998
Genus *Pituna* Costa 1998
Genus *Pterolebias* Garman 1895
Genus *Rachovia* Myers 1927
Genus *Gnatholebias* Costa 1998
Genus *Llanolebias* Hrbek & Taphotn 2008
Genus *Renova* Thomserson & Taphorn 1995
Genus *Simpsonichthys* Carvalho 1959
Genus *Spectrolebias* Costa & Nielsen 1997
Genus *Stenolebias* Costa 1995
Genus *Terranatos* Taphorn & Thomerson 1978
Genus *Trigonectes* Myers 1925

Suborder *Cyprinodontoidei* Wagner 1828
Family *Cyprinodontidae*
Subfamily *Cyprinodontinae*
Genus *Cyprinodon* Lacepede 1803
Genus *Cualac* Miller 1956
Genus *Flordichthys* Hubbs 1926
Genus *Garmanella* Hubbs 1936
Genus *Jordanella* Goode & Bean 1879
Genus *Megupsilon* Miller & Walters 1972
Genus *Aphanius* Nardo 1827
Subgenus *Aphanius* Nardo 1827
Subgenus *Anatolichthys* Kosswig & Sozer 1945
Subgenus *Turkichthys* Ermin 1946
Subgenus *Kosswigichthys* Sozer 1942
Subgenus *Tellia* Gervais 1863

Genus *Orestias* Valenciennes 1839
Subfamily *Cubanichthyinae* Parenti 1981
Genus *Cubanichthys* Hubbs 1926
Genus *Chriopoides* Fowler 1939
Genus *Yssolebias* Huber 2012
Family *Fundulidae* Gunther 1866
Genus *Fundulus* Lacepede 1803
Subgenus *Fundulus* s.s. Lacepeded 1803
Subgenus *Fontinus* Jordan & Evermann 1896
Subgenus *Plancterus* Garman 1895
Subgenus *Wileyichthys* Ghedotti & Davis 2013
Subgenus *Xenisma* Jordan & Copeland 1877
Subgenus *Zygonectes* Agassiz 1854
Genus *Leptolucanis* Myers 1924
Genus *Lucania* Girard 1859
Family *Goodeidae* Jordan & Gilbert 1883
Subfamily *Empetrichthyinae* Jordan, Evermann & Clark 1930
Genus *Empetrichthys* Gilbert 1893
Genus *Crenichthys* Hubbs 1932
Family *Profundulidae* Hoedeman & Bronner 1951
Genus *Profundulus* Hubbs 1924
Subgenus *Profundulus* s.s. Hubbs 1924
Subgebus *Tloloc* Alvarez & Carranza 1951
Family *Valencioidea* Parenti 1981
Genus *Valencia* Myers 1928
Family *Poeciliidae* Bonaparte 1831
Subfamily *Poeciliinae* Bonaparte 1831
Genus *Tomeurus* Eigenmann 1909
Subfamily *Aplocheilichthyinae* Myers 1928
Genus *Aplocheilichthys* Bleeker 1852
Genus *Procotopus* Boulenger 1904
Genus *Aapticheilichthys* Huber 2011
Genus *Hylopanchax* Poll & Lambert 1965
Genus *Hypsopanchax* Myers 1924
Genus *Lamprichthys* Regan 1911
Genus *Patanodon* Myers 1955
Genus *Plataplcheius* Ahl 1928
Genus *Platypanchax* Ahl 1928
Genus *Rhexipanchax* Huber 1999
Genus *Micropanchax* Myers 1924
Genus *Laciris* Huber 1981
Genus *Lacustricola* Myers 1924
Subgenus *Lacustricola* Myers 1924
Subgenus *Cynopanchax*
Genus *Poropanchax* Clausen 1967
Subgenus *Poropanchax* Clausen 1967
Subgenus *Congopanchax* Poll 1971
Genus *Fluviphylax* Whitley 1965
Gamily *Anablepsidae* Bonaparte 1831
Subfamily *Anablepsinae* Bonaparte 1831
Subfamily *Oxyzygonetinae* Parenti 1981
Genus *Oxyzygonectes* Fowler 1916

References

Print Volumes

American Killifish Assc.	Journal of the American Killifish Association Volumes 1961- 2015
Boulenger, G.A. 1969	Catalogue of the Freshwater Fishes of Africa
Boschi, Enrigue, 1957	Argentine Pearl Fish
Brousseau, Roger Dr. 2002	South American Annual Killifish
Hellner, Steffen 1990	Killifish A Compete Pet Owners Manual
Jubb, R.A. 1981	Nothobranchius
Langton, Roger	Breeding Killifish
Neumann, Werner 2003	Pike-like Killies
Ostrow, Marshall 1981	Breeding Killifishes
Scheel, Jorgen 1968	Rivulins of the Old World
Seegars, Lothar Dr. 2000	Aqualog: Killifishes of the World, New World Killis
Seegars, Lothar Dr. 2000	Aqualog: Killifishes of the World, Old World Killis I
Seegars, Lothar Dr. 2000	Aqualog: Killifishes of the World, Old World Killis II
Seegars. Lothar Dr. 1985	Preachtgrundkarpflinge: Die Gattung Nothobranchius
Turner, Bruce, Pafenyk, John	Enjoy Your Killifish
Warner, Edward 1977	Success with Killifish
Wildekamp, Rudolph H.	A World of Killies Vol. II 1995
Wildekamp, Rudolph H.	A World of Killies Vol. III i996
Wildekamp, Rudolph H.	A World of Killies Vol. IV 2004
Windelov, Holger 1987	Aquarium Plants: A Complete Introduction

Online Reference Websites

American Killifish Association	http://www.aka.org
Aquapress Bleher	http://www.aquapress-bleher.com/
Aqua International Journal of Ichthyology	http://www.aqua-aquapress.com/
Bonn Zoological Bulletin	http://www.zoologicalbulletin.de/
Copeia	http://www.bioone.org/loi/cope
Evolution: International Journal of Organic Evolution	http://onlinelibrary.wiley.com
Ichtological Explorations of Freshwaters	http://www.pfeil-verlag.de/04biol/d9902.php
It rains Fishes	http://www.itrainsfishes.net/content/
Killi-Data Online	http://www.killi-data.org/
KillieNutz Online	http://killienutz.com
Nothobranchius Maintenance Group	http://nothos.org/
Senckenberg: World of Biodiversity	http://www.vertebrate-zoology.de/
Wikipedia - The Free Library	https://en.wikipedia.org
Zootaxa	http://www.mapress.com/zootaxa/taxa/Pisces.html

Where you can find Killifish

There are dozens of killifish societies worldwide. Listed below are some of the websites where they can be found.

AKA American Killifish Association:	http://www.aka.org
BKA British Killifish Association:	http://www.bka.webeden.co.uk
DKG German Killifish Group:	http://www.killi.org
APK Assosicao Portuese de Killifilia (Portugal)	http://www.apk.pt/
BKV Belgische Killifish Vereniging (Belgium)	http://www.killiadictos.com/
KFN Killifish Nederland (Netherland)	http://www.killifishnederland.nl
CZKA Czech Killifish Association (Czech)	
CZKZ Czech Killifish Club (Czech)	
SEK Sociedad Española de Killis (Spain)	http://www.sekweb.org/
SKS Skandinaviska Killi Sällskapet (Scandinavia)	http://www.killi.dk/
Belgium (French) Killifish Club	http://www.akfb.be
K.A.B. Killi Association Bulgaria	https://www.facebook.com/KilliAssociation Bulgariametchmdm@bitex.com
K.C.A. Killi Club Argentino 2002	http://www.killiclub.org
K.C.F. Killi Club de France	http://www.killiclubdefrance.org
K.C.J. Killifish Club of Japan	http://kcj.cside.com
S.A.K.S. South African Killifish Society	http://tgenade.freeshell.org
S.E.K. Sociedad Española de Killis	http://www.sekweb.org

Clubs in the United States

Arizona Rivulin Keepers	http://www.aka.org/ark/
Chesapeake Killifish Association	http://www.chesapeakekillifish.org/
Chicago Killifish Association	http://www.chicagokillifish.com/
Keystone Killiy Group	http://www.keystonekilly.org/
Metropolitan Area Killifish Association	http://www.aka.org/maka/
Michigan Killifish Association	http://www.aka.org/mka/
Minnesota Killie Keepers Association	http://www.aka.org/mka/
Northwest Killies	http://www.nwkillies.org/
Pittsburg Area Killifish Association	http://sheneskillies.com/paka/paka.htm
St. Louis Area Killifish Association	http://www.inkmkr.com/Fish/
Southern California Killifish Association	http://www.socalkilliclub.com/
Southern New England Killifish Association	http://sneka.org/
Southern Ontario Killifish Association	https://groups.yahoo.com/neo/groups /Sokskillies/info
Texas Area Killifish Assocition	http://www.aka.org/tako/
Upstate New York Killifish Association	http://unyka.org
Wisconsin Area Killifish Association	http://www.aka.org/wako/